1977 onwards

COVER CUTAWAY: JCB Mk III backhoe loader.

© Julian Carder 2016

All rights reserved. No part of this publication may be reproduced or stored in a retrieval system or transmitted, in any form or by any means, electronic, mechanical, photocopying, recording or otherwise, without prior permission in writing from Haynes Publishing.

First published in September 2016

A catalogue record for this book is available from the British Library.

ISBN 978 1 78521 072 3

Library of Congress control no. 2016930191

Published by Haynes Publishing,
Sparkford, Yeovil,
Somerset BA22 7JJ, UK.
Tel: 01963 440635
Int. tel: +44 1963 440635
Website: www.haynes.com

Haynes North America Inc.,
861 Lawrence Drive, Newbury Park,
California 91320, USA.

Printed in Malaysia.

WARNING
Whilst every attempt has been made throughout this book to emphasise the safety aspects of working on, restoring and operating a JCB 3C Mk III backhoe loader, the author and publishers accept no liability whatsoever for any damage, injury or loss resulting from the use of this book.

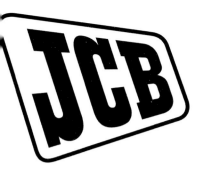

JCB 3C MkIII BACKHOE LOADER

977 onwards

Owners' Workshop Manual

An insight into the design, engineering, maintenance and operation of JCB's iconic backhoe loader

Julian Carder

Contents

6	Foreword	
7	Introduction	
8	JCB: one of Britain's great success stories	
	J.C. Bamford Excavators Ltd	11
	Tipping trailer	12
	Major Loader	13
	Rocester factory	14
	Master Loader	15
	Si-draulic Loader	15
	Half-Track	16
	Hydraulic Excavator Trencher	16
18	Development of the Backhoe Loader	
	Mk I Excavator	20
	Loadall power shovel	21
	Hydra-Digga	21
	JCB 4	23
	JCB 4C	24
	JCB 3	24
	Hydraslide	25
	International track loaders	26
	JCB 3C	26
	JCB 1	26
	JCB 2	27
	JCB 2B and 2D	28
	JCB 3D	28
	JCB 4D	28
	Mk II range	29
	Leyland 4-98 skid unit	30
	Mk III range	30
32	Manufacture and sales	
	£250,000 staff bonus	34
	Queen's Award to Industry	35
	Market leader in 52 countries	35
	Manufacturing efficiency	36
	World demand	36
38	Anatomy of the JCB 3C Mk III	
	Hydrachassis	40
	Engine	41
	Transmission and gearbox	45
	Brakes	48
	Excavator	49
	Loader	51
	Front axle	52
	Wheels and tyres	53
	Power steering	54
	Hydraulics	55
	Cab	60
	Controls	62
	Seat	63
	Electrics	64
	Instrumentation	65
	Optional and additional equipment	65
	JCB Roadbreakers	67
	Air conditioning	69
	Cold start	70
	Extending dipper	70

BELOW 2016 marks the centenary of the birth of JCB's Founder Joseph Cyril Bamford CBE. He was born on 21st June, 1916 at the family home, The Parks, Uttoxeter.

ABOVE **JCB 3C Mk III.**

| 72 | The designer's view |

Torque and power	74
Efficient design	74
Pins and bushes	74
Performance	75
Excavator design	76
Boom and dipper design	76
Loader design	77
Testing the product	78
Part numbers	81

| 82 | The owner's view |

Choosing the right specification	84
Where to look for a machine	85
What to look for when buying	85
Restoring a machine	88
Care and safety	92
Routine maintenance	93

| 104 | The mechanic's view |

Safety first	106
Dismantling and assembly	106
Bearings	107
Servicing	107
Component removal	110
Electrics	112
Troubleshooting	113
Lubricants	113

| 114 | Marketing |

The JCB logo	116
Advertising	116
Merchandise	120

| 126 | Epilogue |

| 128 | Appendices |

A: Pricing	128
B: 3C development	129
C: Attachments	137
D: Machine identification and serial numbers	146
E: Specifications	147
F: Export standard builds specifications	150

| 151 | Useful contacts |

| 154 | Index |

Foreword

by Lord Bamford

My first memory of JCB really was my father and it's true to say he was an engineering genius. He was always dreaming of better ways of doing things. When he invented the backhoe loader in 1953 the product really was a godsend and it genuinely started a mini revolution in construction machinery. He could never have imagined how successful it would become over the following six decades. Over that time, JCB has manufactured more than half a million backhoes and today it remains one of the world's most versatile pieces of construction machinery. While the past is the past and we are very much focused on developing the innovative backhoes of the future, looking back into history can be a source of enormous pleasure. For that reason I'm delighted that Haynes is publishing this unique manual on one of JCB's most iconic models: the 3C Mk III. I really hope you enjoy reading it as much as I did.

Antony Bamford.

Introduction

In the immediate post-war period, the United Kingdom, in common with most of Europe, had an urgent need for rebuilding work. Rubble and damaged structures needed to be removed before building could begin, but work was hampered by a desperate shortage of labour and equipment.

The plant-hire industry was carried along by this post-war boom, and the plant hirers – who had often learned their trade and increased their efficiency during the war years – enjoyed busy times and a full workload. A shortage of machinery, due to manufacturers' inability to deliver, was the only factor which stopped the industry expanding further, and many popular items of plant were only available on a quota basis.

Unlike the contractors, who chose machinery designed for a specific job or application, plant-hire companies needed machines that were versatile and able to tackle various tasks on different sites. A demand therefore arose for a multipurpose machine, capable of moving readily from one site to the next and which could be operated by a relatively unskilled labour force.

The only earthmoving machines available at the time were designed for specific purposes, such as bulldozers and rope shovels. The loading shovel had a winch-driven hoist to raise the bucket, a frame, and a mechanical trip to discharge the load. Dozer blades were also adjusted by cables. Hydraulics were hardly used, except on major pieces of plant.

Early attempts to produce more versatile machines were made by modifying agricultural tractors to carry a loading shovel, an excavator arm (or backhoe), or both. These machines, however, were only effective up to a point; their success was limited because they were never designed to take the severe stresses and loads involved with an earthmoving machine. Nevertheless, the concept proved sound and capable of further development.

The JCB approach was to produce such a vehicle specifically designed for earthmoving. This immediately gave the market what it wanted – a practical machine at a reasonable cost that could perform well and was stable, easy to operate, manoeuvrable and, above all, reliable.

The most important piece of plant for the hire industry was to become the backhoe loader, which could never have been conceived in the days of steam, winch, lever and cable. The versatility of these hydraulic machines appealed to the plant-hire firms, as they were small enough for the average building site and had the added asset of mobility, being able to travel from yard to site without the use of a low loader.

LEFT **Before hydraulics, machines were operated by ropes and pulleys, as on this early Chaseside loading shovel. JCB acquired Lancashire company Chaseside Engineering in 1968, and started to produce wheeled loading shovels.**

Chapter One

JCB: one of Britain's great success stories

JCB – the world's number one backhoe loader manufacturer – is a global engineering company which celebrates its 70th Anniversary in 2016. The company is still privately-owned by the Bamford family, and the current Chairman, Lord Bamford, has been at the helm since 1975.

The foundation for the growth and success that was to follow was the manufacture of a tipping trailer made from war-time scrap.

OPPOSITE **The JCB Dancing Diggers perform their 'up and under' manoeuvre during a coordinated routine in JCB's demonstration yard at Rocester.**

RIGHT JCB founder – Joseph Cyril Bamford CBE.

JCB's founder, Joseph Cyril Bamford, came from a family that began manufacturing agricultural equipment as Bamfords Ltd in 1879, having started as blacksmiths in the 1830s based in the market town of Uttoxeter in Staffordshire, England.

Joseph was the great-grandson of Henry Bamford who, on his marriage to Julia Brassington in 1844, opened an ironmongers' shop in Uttoxeter market place. Henry and Julia had 11 children, the eldest of whom was Samuel (1845–1932). Samuel had an inventive mind, and after leaving school he started to develop products outside ironmongery. He made pumps and galvanised troughs and patented such things as iron taps for water butts. He persuaded his father to purchase some land on which to build a small foundry, which later became the site of Leighton Iron Works and subsequently the home of Bamfords of Uttoxeter.

Samuel's brothers all played a part in the growing family business, and of these it was Joseph Bamford who was sent to run a new family concern, the Lichfield Agricultural Company, which sold the expanding range of products manufactured at the Uttoxeter factory.

Bamfords of Uttoxeter developed many labour-saving machines for use on farms. The company later developed stationary engines to

RIGHT Mr JCB and his wife, Marjorie, with their two sons, Anthony (left), now Lord Bamford, and Mark.

drive these farm machines, thereby improving productivity. At this time there wasn't a farm in the country that didn't have a Bamfords of Uttoxeter product.

Cyril Joseph married Dolores Turner and they lived at 'The Parks' in Uttoxeter. They had five children, the eldest of whom was Joseph Cyril Bamford ('Mr JCB'), who in 1945 set up his own business – JCB.

ABOVE The Bamford factory production line. The smaller grinding mills are hand assembled on a separate line to the larger versions in the background.

BELOW The rented garage in Uttoxeter in which it all began. Mr Bamford's first machine tool purchase was a prototype welding set for £1 from English Electric. The garage was demolished in 1969 and a replica now stands at Lakeside Works at Rocester.

J.C. Bamford Excavators Ltd

J.C. Bamford Excavators Ltd was founded on 23 October 1945, when Joseph Cyril Bamford, then aged 29, decided to break away from the family firm and go into business on his own, beginning the company that would take his initials around the world.

At first he couldn't afford to buy anything and was paying 30 shillings (£1.50) a week rent for a lock-up garage in Derby Road, Uttoxeter, with a working area of approximately 12 by 15 feet (3.6m by 4.5m). There was no electricity at first, because not everyone was allowed to connect

ABOVE Mr JCB next to the hydraulic tipping trailer made from wartime scrap at the factory in Crakemarsh Hall near Uttoxeter.

with the mains supply system for industrial purposes while everything remained rationed after the war. His only assets were a second-hand English Electric welding set, which had cost a pound, an engineering background and a determination to succeed.

On the same day as the company was founded, Joseph Cyril's son Anthony – now Lord Bamford – was born and, as Mr Bamford remarked, 'Being presented with a son tended to concentrate the mind, and when you were starting at the bottom there was only one way to go, and that was up.'

Tipping trailer

Mr Bamford's first product was a screw-tipping farm trailer made mainly from war-surplus materials. The material for the body panels was salvaged from Anderson air-raid shelters, which was corrugated and had to be rolled flat, while the tyres and axles were taken from old Albion trucks. After the trailer was painted with coachlines and his signature was added he took it to Uttoxeter market and advertised it for sale for £90. Eventually, on the third Wednesday of showing, after nearly every farmer in the area had looked at it and supposedly found something wrong, it was sold for £45, with the buyer's old cart taken in part exchange. Having renovated the farm cart, adding a fresh coat of new Brunswick green paint and some coachlines, Mr Bamford returned to Uttoxeter market the following week and sold it for another £45, thus achieving the original asking price of £90.

Mr Bamford handled almost every job on the trailers himself, including cutting, forming and welding. Unable to afford a drilling machine, he contracted out all the drilling work to a local bus garage. He even hand-painted the trailers himself for a time until he could afford to employ someone on a part-time basis in the evenings.

In mid-1946 the company's first four-wheeled

BELOW 1947: This picture shows Mr Joe Bamford with two-year-old son Anthony and (from left) Bill Hirst MBE, who became director; Arthur Harrison, JCB's first foreman, who ran the machine shop; and Bert Holmes, the company's first welder. In the background is one of the company's all-steel tipping trailers. The wheel and tyres are from a Grumman Hellcat fighter aircraft, the hubs from a small howitzer field gun.

ABOVE 1948: John Wheeldon became employee number six and went on to be in charge of the first JCB-operator training school. He is pictured here with parts from a Major Loader outside the stable block at Crakemarsh Hall.

trailer was built, and Mr Bamford started refurbishing wartime Jeeps, which were in high demand, by adding wooden bodies supplied by a local undertaker. This proved to be very profitable due to vans and cars still being in short supply after the war.

By 1947 the company was expanding and had outgrown its original cramped facilities. As Mr Bamford's landlady also disapproved of his Sunday working, he moved a few miles down the road to a stable block at Crakemarsh Hall, which was owned by a Mrs Julia Cavendish, a survivor of the *Titanic* disaster. Mr Bamford also took on his first-ever full-time employee, Arthur Harrison, who later became foreman.

The company's first tipping trailer had a manually operated screw device. The first JCB machine to utilise a hydraulic system as a means of movement was an agricultural trailer. Produced in 1948, it was Europe's – and possibly the world's – first two-wheeled, hydraulic, all-steel tipping trailer. This made life much easier for the operator during the tipping operation, but after it had tipped it was then necessary to drive halfway around the field in order to make the trailer bed lower again, as the double-acting ram mechanism was yet to be devised.

Major Loader

At this time farmers were still loading trailers mainly by hand, and it was for this reason that thoughts turned to utilising machinery. In 1948 Mr Bamford used his knowledge of hydraulics in aircraft landing gear to set about designing a hydraulic loader attachment to mount on tractors, employing the hydraulic ram technology he had developed for his tipping trailers.

The company's 'Major Loader' was the first hydraulic loader in Europe. Introduced in 1949, it was designed to be mounted on the Fordson Major – the most popular tractor of the time – and Morris Motors' Nuffield Universal tractors. It was a well-built loader, fitted with two powerful, single-acting hydraulic-lift rams of patented design, and had a maximum working height of 3.9m (12ft 9in). The price was around £150.

BELOW 1948: The Major Loader name came from the fact that it could be bolted on to a Fordson Major tractor, although it could be fitted on to other tractors too. It is seen here on a Nuffield driven by John Wheeldon.

ABOVE 1948: Bill Hirst MBE using an angle grinder to clean up the welds on the shovel for a Major Loader, watched by Herbert Tickhill.

so they could be renewed, a general-purpose shovel suitable for loading loose material, a crane jib that could lift up to 762kg (1,679lb), and a light bulldozer blade suitable for site clearing and back-filling trenches.

The later 'High Speed Major Loader' was significant because it marked the company's first move into the industrial and construction markets, and was a distinct departure from the agricultural equipment that had been JCB's initial market. With the High Speed, JCB was also beginning to move away from tractor hydraulics. It was fitted with a pump, a valve block and a separate circuit and hydraulic tank bolted to the side of the frame. This is when it began to take off as a loading shovel rather than a farm implement, because customers realised it could handle tough work.

The Major Loader remained in production until March 1957, when it was replaced by the JCB Loadall.

Rocester factory

Sold only in kit form for the dealer or farmer to bolt on to his own selected tractor, part of the loader was made from ex-air-raid shelter material, mainly four-by-two-inch channels. It was later developed for mounting on all popular makes of tractor, and thousands were sold. Optional attachments for the Major Loader included a muck fork with detachable tines

By 1950 the business had outgrown the stables of Crakemarsh Hall, and JCB purchased and moved into the old Wiltshire United Dairies milk processing and cheese factory at Rocester,

RIGHT Mr Bamford was on the lookout for new premises but was keen to remain close to his home in Uttoxeter. The site eventually chosen was spotted by Bill Hirst MBE, as it was close to his own home in nearby Rocester. It was a former cheese factory with just under an acre (0.4Ha) of buildings and land. The JCB HQ is still at Rocester, although the site now stretches to 175 acres (70Ha).

which had also been a pig farm and latterly a cattle market. This has since become the company's world headquarters.

Mr Bamford's office at the new Lakeside Works, as it was renamed, was converted from a deep litter chicken run. Over three inches of chicken droppings had to be cleared from the floor during construction, and the smell lingered for months!

Master Loader

In 1951 the 'Master Loader' was introduced for the Fordson E27N tractor. A smaller, less powerful version of the Major Loader, this was specifically aimed at farm work and was ideally suited to loading muck into farm carts, trailers and muck spreaders, which were gaining popularity at the time. The selling price for the completed machine was £75, and around 2,000 were built.

Si-draulic Loader

1953 was a turning point for JCB, with a world first in the shape of the innovative Si-draulic (single-arm) loader. Designed to bolt on to the axle and supported by suitable braces for Fordson tractors, it was also available for most other leading tractors of the time. Its single loader arm was made of 114mm (4½in) tube braced with a tie rod and it had a maximum lift height of 3.4m (11ft). It was to remain for many years the only loader mounted on one side of a tractor. It offered high lift and good forward reach unattainable by any of its competitors. At

LEFT 1951: The Master Loader was a smaller version of the Major Loader. Though this sales leaflet predates the creation of JCB's own logo, the name 'J.C. Bamford' is very boldly displayed at the bottom. Underneath it states 'Our only address, no connection with any other firm', as Joe Bamford didn't want customers to think he was part of Bamfords, the family business which he had left.

BELOW 1953: The Si-draulic loader was a single-arm loader unit that was a simple bolt-on kit for tractors and sold for £75.

£75 it once again represented extremely good value for money. Some 6,000 of these units were produced by JCB and a further 20,000 units were produced under licence in France. The arm could be detached from the tractor in about ten minutes, which left the rear of the tractor and the drawbar clear for operating other hydraulic implements.

Half-Track

As well as continuing with the development of loader unit derivatives, throughout the early 1950s Mr JCB also produced mid-mounted grass mowers for a wide variety of tractors, before eventually selling this side of the business to Bamfords Ltd of Uttoxeter so that he could concentrate on the production of loaders, which were in high demand.

In 1952, when half-track conversions of agricultural tractors were becoming popular, Mr Bamford designed and built a Half-Track design to aid the traction of vehicles working in deep mud and on snow-covered ground. To complete the conversion, a set of skis was also available to replace the front wheels; an idler axle and wheel were added between the front and rear wheels and the original wheels were retained to drive the rubber belt with steel plates that acted as the track.

These conversions sold for £185 but had little success in the UK, and though more successful in Scandinavia they were discontinued in 1957 as they weren't selling in the large numbers that had been expected. Three tractors fitted with JCB Half-Tracks, complete with front skis, were driven to the South Pole in January 1958 by Sir Edmund Hillary.

Hydraulic Excavator Trencher

It was on a sales trip to Norway to sell the JCB Half-Tracks that Mr JCB saw what would prove the foundation of the company's success for years to come. He noticed something new – a lightweight hydraulic backhoe mounted on wheels for digging trenches. The potential didn't escape him, and even though the machine was very restricted and would only slew to about 80° from the centre line, he bought one of these units and shipped it back to England.

After examination in the workshop, the conclusion was reached that something better could be built by JCB. Lessons learned from evaluating the Norwegian machine were utilised in designing and producing the first JCB backhoe, which was called the 'Hydraulic Excavator Trencher'. The JCB excavator was based on an 'A' frame attached to the tractor's three-point linkage, being raised by the tractor's rear hydraulic lift. The boom and dipper were made of tubular steel and operated via a linkage bar and hydraulic ram. The movement of the

BELOW 1952: A rare sales leaflet promoting the JCB Half-Tracks, showing Ferguson (top) and Fordson Major (bottom) tractors fitted with optional skis in place of the standard front wheels.

RIGHT 1952: The JCB Excavator, or 'Trencher' as it was known, fitted on to the back of a tractor's three-point linkage.

excavator was controlled by five individual levers connected to the hydraulic control valve block, which at the time was supplied by Spenborough Engineering Ltd, along with the hydraulics rams.

The buckets – or scoops, as they were originally titled – were available in 25cm (10in), 45cm (18in) and 60cm (24in) widths. With these the Excavator could dig to a depth of 2.8m (9ft 3in), while having the capability to slew a full 180°.

The Hydraulic Excavator was available to fit on Fordson Major, Nuffield and Ferguson tractors and cost £850 ex-works. Mr Bamford soon realised that a Major Loader could also be fitted to the front, thereby providing an even more versatile tool. This is how Europe's first revolutionary backhoe loader was born.

BELOW This is the company's impressive stand at the UK's Royal Show of 1953.

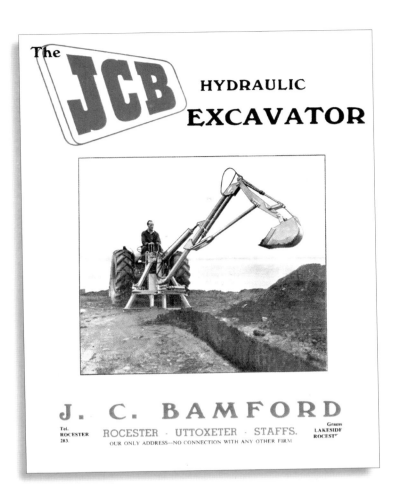

Chapter Two

Development of the Backhoe Loader

In 1953, Joseph Cyril Bamford combined a lightweight backhoe with his Major Loader to create a machine that has quite literally built the world. The genius was in bringing both the loading shovel and rear excavator arm together to produce a highly versatile machine. From that early JCB Mk 1, the concept developed rapidly and, not surprisingly, the Backhoe Loader will always be regarded as JCB's most important innovation, which proved to be the building block for its success.

OPPOSITE The side-by-side comparison of a 3C Mk II and a Mk III makes it very easy to see the latter's taller and more spacious cab.

ABOVE 1953: The Mk I Excavator brought the new wonder of hydraulic power to construction for the first time. The rear bucket could be turned around to use as a face shovel, as seen on this machine owned by Amey's Asphalt of Abingdon, Berkshire.

BELOW 1955: A JCB Mk I Excavator, fitted here with the front counterweight option. Mr Quick, a JCB dealer from Cheshire, is presented with an award for commitment and service.

Mk I Excavator

The Mk I was introduced in 1953 and was an immediate success, giving excellent service to both builders and farmers, especially when fitted with the Major Loader to provide an ideal machine for digging, loading, lifting and levelling. The early booms on the excavator were made from box-section beams and used tubular braces to provide additional strength, but this means of construction was quickly replaced with a heavier-duty design that did away with the need for braces. It also had side panels that were screwed to either side of the boom, giving it a more robust appearance and providing a mounting position for the 'JCB' and 'Excavator' name badges.

The Mk I Excavator came with either a concrete counterweight on the front, an air compressor, or was fitted with the Major Loader with a $0.48m^3$ ($5/8yds^3$) shovel to make it a true backhoe loader, referred to as the 'Mk I Major Loader'. The excavator was fitted to an 'A'-frame design installed on the rear of the tractor, and a large hydraulic ram was used to lift and lower the entire rear excavator. Loader lift-height was 3.65m (12ft), it had a maximum dig depth of 3.3m (11ft), maximum reach of 4.8m (16ft), and could excavate up to $34m^3$ ($45yd^3$) per hour. A choice of four buckets was available as well as a tapered ditching bucket and ditching blade. A face shovel was also later developed, which was marketed as the 'most powerful hydraulic face shovel in Great Britain'. The number welded on to the side of the bucket-tipping link was that which was allocated to the welder who produced that particular Excavator.

The Mk I Excavator and Major Loader were available both as a kit for dealers to fit on to new or used tractors, and as a finished machine from the JCB factory. The colour scheme for dealer-fitted machines was Fordson empire blue tractor with Fordson orange wheels and cardinal red excavator and loader. Factory-assembled machines had a yellow tractor unit and red wheels, loader and excavator.

LEFT 1956: The JCB Mk I Major Loader on the company Bedford RL low-loader demonstration vehicle, with a set of rear buckets and the bulldozer blade fitted to the front.

Loadall power shovel

Mr Bamford's 'excavator empire' was growing rapidly due to the success of the Mk I Major Loader, which continued to sell well to both builders and farmers, particularly when fitted with the Major Loader to provide the ideal machine for digging, loading, lifting and levelling.

However, Mr JCB wasn't satisfied with certain elements of its design, such as the externally mounted hydraulic tank, and decided that in order to keep it competitive and ahead of its competition it needed to be strengthened. He therefore set about designing a replacement.

In 1957 the Loadall power shovel was introduced for the Fordson Major, Nuffield, Ferguson and other tractors. For the first time it utilised JCB's innovative Hydrachassis hydraulic system, where the oil was drawn from the loader's hollow mounting frame, then to the rams and finally back to the separate hydraulic tank via two oil filters. The Loadall's front shovel was hydraulically tipped and crowded via a double-acting ram, unlike the JCB Major Loader, which had a wire trip mechanism. The deep, wide arms were de-bossed to give extra strength and braced by box pressings to form a rigid structure. In 1958 the Loadall was renamed the Loadall 65, and a more powerful version, the Loadall 75, was introduced which featured large-diameter hydraulic rams and a higher-flow gear pump. The Loadall 65 had a lift capacity of 1,524kg (3,360lb), compared to 2,495kg (5,500lb) on the more powerful Loadall 75.

Hydra-Digga

The Hydra-Digga excavator was launched in 1956. It wasn't a replacement for the Mk I Excavator but was sold as a larger, more powerful version. The boom and dipper were larger and stronger and had larger-capacity excavator buckets, but the 'A'-frame mounting was similar to the previous Mk I. Its name,

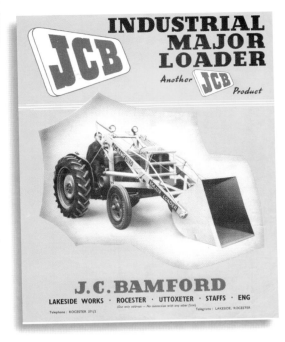

LEFT Though JCB's business was selling to agricultural customers, Mr JCB saw an opportunity for a heavier-duty front-end loader in the construction market, and so marketed his Major Loader as the 'Industrial Major Loader' to promote its strength.

BELOW A prototype Loadall being tested by Mr Joe Bamford in 1955.

RIGHT The Hydra-Digga excavator could be supplied with a Major Loader fitted to the front, as seen on this machine digging a hole for a fuel storage tank at an Esso petrol station circa 1958.

ABOVE 1957: The Hydra-Digga with Loadall 65 loader fitted to the front. The stabilisers' extensions are in place to give increased stability when working to the side, but they did make the machine cumbersome to move around site.

ABOVE RIGHT The Hydra-Digga could also be fitted to a variety of track loaders including International, Track Marshall and County Crawlers, but the added weight on the back made machines too light at the front, which had an adverse effect on their performance.

BELOW The introduction of two levers instead of six for the JCB 4's main controls was a major innovation, reducing operator effort and increasing productivity.

incidentally, was devised by Mrs Bamford from the word 'hydraulics' and the Hydra of Greek mythology (a creature with numerous heads).

The Hydra-Digga was available as a single unit with a front counterweight, or as an Excavator/Loader with the Major Loader fitted and a choice of Fordson or BMC Nuffield tractor unit.

Following the introduction of the Loadall 65 in 1957 the Hydra-Digga and Loadall together constituted the 'Hydra-Digga/Loadall 65' combination – the first purpose-built earthmover with a separate Hydrachassis, which utilised the hollow sections of its chassis to carry the hydraulic oil instead of needing a separate tank. It was also the first model to offer the operator a reasonable amount of comfort in the cab. To help give the impression of strong boom construction it was fitted with a pressed-steel cover to protect the ram, but due to its size this was susceptible to damage and was often removed.

The Hydra-Digga/Loadall 65 was the first JCB machine to have power-assisted steering. It was fitted with the 'Hydrosteer' system, which was operated by the standard tractor steering box and drag link arm coupled to a hydraulic ram.

Although the Hydra-Digga excavator was designed for the Fordson Major tractor it was also available for the International B-450 tracked machine, due to the popularity of tracked dozers and loaders at that time. In addition it was available as a dealer kit for Track Marshall, International BTD6 and County Crawler tractors.

In 1959 a new two-lever excavator valve block for all digging operations was introduced. Previously there were four levers for operating individual rams and one for lifting and lowering the 'A' frame. The combined research teams of JCB and the valve supplier Hamworthy had been developing their revolutionary two-lever principle for some considerable time. It was a key innovation and was said to increase output by 15%, as the operator's hands remained on the two operating levers for the complete work cycle and resulted in faster digging. Over 2,000 Hydra-Diggas were sold before production ceased in 1960.

RIGHT JCB 4 production line in 1960. After the machines had been built they were cleaned and then painted. The Fordson skid units can be seen on racking in the background, ready to be used.

CENTRE In 1960 JCB replaced the still-popular Hydra-Digga Loadall with the JCB 4, which was capable of carrying several people in its large cab.

JCB 4

The success the JCB Hydra-Digga/Loadall 65 had, during 1959, prompted the development of a larger, more powerful machine for civil engineering work and contractors. The JCB 4 was introduced in 1960 to replace the Hydra-Digga/Loadall and was a more sophisticated and slightly more powerful machine than its predecessors. Very popular with public works' contractors, and used on motorway projects throughout the 1960s, this new machine was initially built around a Fordson Power Major until the Super Major skid unit was introduced. It incorporated an updated Loadall-type front shovel and built-in mudguards carrying the hydraulic oil reservoir, plus a 'Spaceview' cab large enough to seat five people, which was marketed as both site-to-site transport and a site office! Another new feature of the JCB 4 was a specially built front axle and yoke, which had a certified rating of 70 tonnes to withstand the heaviest shock loading. A 4.8m-wide (16ft) 'A' frame provided stability when digging and had a reach of 6.7m (22ft) with extension arm fitted and 5.8m (19ft) without, which were unequalled by any other machine.

The JCB 4 had larger-capacity buckets than the Hydra-Digga, due mainly to the larger bucket ram that was fitted, and had the new two-lever digging controls first introduced on the Hydra-Digga in 1959. It was also the first JCB to offer a fuel filling point outside of the cab, due to the use of a JCB fuel tank instead of the standard Fordson tractor version.

In addition, in 1962 the JCB 4 was the first machine to demonstrate the company's 'up and over' routine, in which a car was driven underneath a machine elevated on its hydraulic rams. This marked the beginning of the world-famous 'Dancing Diggers' troupe, which gave its first official performance two years later.

BELOW In 1962, the JCB 4 was the first JCB machine to perform the 'up and under routine', where it was elevated on its own hydraulic arms while a car – in this photograph the company Cadillac – was driven underneath. Company demonstrator Noel Hooper came up with the idea for the stunt, for which he was awarded £5 by Mr Bamford.

RIGHT The JCB 4 was updated to the 4C in 1962 with an improved cab and squared fender design. In this photograph, taken opposite the JCB Rocester factory in 1964, a 4C is being demonstrated by Henry Finney loading a JCB 1D site dumper.

JCB 4C

In 1962 the JCB 4 was updated to become the JCB 4C by the fitment of square mudguards, a new-style cab and longer-reach front shovel arms to improve loading over the sides of lorries. It was powered by the 52hp Fordson Super Major coupled to a Vickers intra-vane-type hydraulic pump with an output of 136 litres per minute (30gal/min). The front loader arms were redesigned to increase loading height, and twin shovel-tipping rams were fitted to give balanced power and to improve access to the engine and hydraulic pump. The excavator was also redesigned to increase loading height by an additional 300mm (12in) to 3.5m (11ft 6in), which enabled the loading of high-sided vehicles. Other improvements included a single-lever valve block, raised driving seat and controls to improve visibility, and a level platform in the cab. The excavator buckets were also modified to allow considerably more crowd action, and a long dipper option was available instead of a dipper arm extension, which increased the reach of the excavator by 914mm (3ft). The 4C was priced at £3,045 ex-works, which included as standard the 'Superview' cab with working lights and fitment of both hand and foot throttles.

JCB 3

Though the JCB 4 had been an immediate success it proved too large for many smaller builders and contractors, and had difficulty competing against the smaller Massey Ferguson and Whitlock machines that were now being used by some companies. This left a gap in the market for a smaller JCB.

In the next few years many other competitors were to appear, endeavouring to take a share of what was becoming a major construction plant market. Earthmoving companies with now forgotten names, such as Inchgreen, Steelfab, Whitlock and Wain-Roy, rose and fell as the ferocity of the marketplace caused them either to fall by the wayside or to manufacture alternative products.

BELOW The JCB 3, launched in 1961, was smaller than the JCB 4 and was consequently marketed on its compactness.

After many years of research and development, the JCB 3 was previewed at the first International Construction Equipment Exhibition held at Crystal Palace, London, in June 1961. Priced at £2,500, the machine received a great deal of interest, and provisional spec sheets were given out to prospective customers. Production started in November 1961 and in 1962 underwent a series of changes, including squaring off the mudguards instead of using round wheel arches and redesigning the excavator kingpost carriage and kingpost to simplify access to and maintenance of the hydraulic hoses. The rear pillars of the cab were also reduced in thickness to improve visibility when the excavator boom was slewed round by 90°. Later changes included a redesigned square-profiled bonnet with shorter sides for greater rigidity, easier engine access and improved engine cooling. A new pump guard provided easier access and more robust protection for the radiator and pump.

Compared to the JCB 4C, the JCB 3 was a smaller, lighter and even more manoeuvrable machine powered by the 52hp Fordson Super Major diesel engine supplied from Ford specially fitted with a petrol-engine-type radiator and six-blade pusher fan for improved cooling. Hydraulic power was derived from a Vickers intra-vane-type pump delivering 94.5 litres per minute (21gal/min) capacity. System pressure was 121 bar (1,750psi). The Hydrachassis box-section mainframe acted as the hydraulic oil tank.

Hydraslide

The key innovation of the JCB 3 was its Hydraslide rear kingpost, which could be moved into any position on the 1.73m-wide

ABOVE Aimed at the small builder, the JCB 3 featured the innovative Hydraslide system that enabled it to be positioned along the rear frame and to dig close to walls.

BELOW 1962: The JCB 3 model had its own dedicated production line while the JCB 4s were built on a separate line, which can just be seen in the background.

BELOW A JCB 3 being driven out of the paint shop in 1963 to have its glass and front shovel fitted.

BOTTOM An advert from *Contract Journal* in 1964, showing three JCB 3Cs working on the A50 Sandacre bypass.

(5ft 8in) frame. Previous models hadn't had this functionality due to their fixed centre-mount positioned kingpost. The JCB 3 also featured, for the first time, independent hydraulic stabilisers vertically positioned at either end of the rear frame, which could level up the excavator on ground sloping at up to 10°. Its offset digging capability made this machine ideal for trenching close to walls or scaffolding and ditch cleaning. Moving the kingpost required the four large nuts to be slackened; the excavator boom was then used to push the carriage across.

International track loaders

Keen to exploit the large track-crawler loader market in the UK, JCB also offered a factory-fitted package to add the JCB 3 excavator unit to both new and used International BTD6, BTD8 and BD-264 machines. However, although the addition of the backhoe increased versatility it added considerable weight at the rear, which in turn made it light at the front. This compromised dozing and loading performance and the package consequently enjoyed only limited success.

JCB 3C

In 1963 JCB identified an opportunity for a more powerful version of the JCB 3 and introduced the 3C, a fully integrated backhoe built around a large mainframe, the JCB Hydrachassis. The 3C, a purpose-built backhoe loader, was the logical extension of the well-established JCB 3, and, by the end of the year, it had already become the highest-selling backhoe loader in the UK.

It was a larger but otherwise almost identical version of the JCB 3, with longer-reach front shovel arms, a larger capacity front shovel, a longer dipper arm and a more spacious cab. Initially powered by the same Fordson Super Major tractor skid as the JCB 3 and 4C (subsequently changed to Nuffield and then to BMC/Leyland, as explained under the 'JCB 2' heading opposite), it had a maximum reach of 18ft (5.49m) and maximum digging depth with extension piece of 13ft (4.06m). It was priced at £2,800, compared to the JCB 3's £2,519.

JCB 1

Market research had meanwhile shown that there was a need for an excavator suited to small builders, which could be used for trenching and all excavating jobs in situations normally inaccessible to larger machines. Six pre-production machines were built based on the JCB 1D dumper chassis which was also being produced at the time, but the production machines had their own integrated chassis.

RIGHT JCB 1 was aimed at small builders, but proved popular with local authorities for cemetery use, and was more commonly known as the 'Grave Digger'.

The JCB 1 was launched at the Public Works & Municipalities Exhibition at Olympia, London, in November 1962, at a selling price of £1,250, with an optional £79 all-weather protective 'Spaceview' cab with a side-opening door. It was powered by a 20hp Petter PH2 twin-cylinder air-cooled engine, with optional electric starting instead of hand-start and fitted with a three-speed and reverse gearbox and an ENV double-reduction rear axle. It had a bulldozer blade at the front rather than a shovel for backfilling trenches, and the centrally mounted fixed excavator had a 3.4m (11ft 4½in) reach and could dig to 2.29m (7ft 6in) depth. An optional 300mm (12in) longer dipper that provided additional reach and dig depth was priced at £47.

It was a lightweight machine weighing just 2,809kg (6,191lb), which made it ideal for working on grass. Many were sold to local authorities for use in cemeteries, which led to it becoming known as the 'Grave Digger', but sales were limited at just 310 machines. Criticised for being too small, production finally ceased in May 1971.

JCB 2

In 1963 JCB started to test the prototype for the JCB 2, which was introduced in 1964 to plug a hole in the company's range between the JCB 1 and JCB 3. When launched it was fitted with the Fordson Major tractor, which was then replaced with the BMC (Nuffield) 342 three-cylinder 42hp skid, which in 1965 was replaced in turn by the 55hp BMC engine with its ten-speed gearbox.

Advances in the design of ram geometry meant that it was now possible to introduce a measure of parts standardisation within the company's range of products. This was reflected in the interchangeable rams found on the JCB 2 and JCB 3, which reduced the variety of spares required for customer fleet operation. The front tyres were 7.50 x 16, which

LEFT Due to its compact size the JCB 1 could work in confined spaces that were inaccessible for larger machines. Today these machines are quite sought-after by collectors, as they're light enough to be towed on a trailer to shows.

BELOW The JCB 2 was introduced in 1963 and was smaller and lighter than the JCB 3.

RIGHT A JCB 2B in 1965, digging a trench at the side of a road for a new sewer.

was a common size also used on Land Rovers, while the rears were six-ply 13 x 24. Fitted with a smaller cab than the JCB 3, the JCB 2 weighed just over 5,000kg (11,000lb) and had an excavating depth of 3m (10ft), a reach of 4.5m (14ft 7in) and a loading height of 2.8m (9ft 10in).

JCB 2B and 2D

The JCB 2B was introduced in 1964. It had a detachable rear excavator with the option of full tractor hydraulics, and three-point linkage incorporating a power take-off and swinging implement drawbar, enabling it to be used for normal farm work. It was primarily aimed at local council and farm contractor use, but it was also popular with electricity boards, as they could fit the JCB pole-planting rig on the back of the machine, allowing one man to plant a pole. A number of other three-point hitch attachments were marketed for local authority use, including a snow blower, a hydraulic grader for levelling work, a road-sweeper collector, a flail mower and a choice of two different Atlas Copco compressor units to power a variety of air tools.

The 2B Mk II was introduced in 1968 with an upgrade to the styling and a larger, more spacious cab with an elevated seating position to improve operator visibility of both the front shovel and the excavator. A major development was the introduction of full power steering, which reduced operator effort and fatigue and was a more robust design than the Hydrosteer power-assisted system that had been used previously.

In 1971, the BMC Nuffield 3/45 tractor skid engine was replaced with the Leyland 70hp 384 skid, as fitted to the rest of the backhoe loader range, and the machine was renamed the 2D.

JCB 3D

During 1967 the JCB 3D was introduced. This was both larger than the 3C and shared the Leyland 70hp engine. The 3D was marked as a de luxe model, came fitted with a 'Hydraslide' kingpost as standard (see page 25) and could dig to 4.6m (15ft 9in). A two-position dipper offered a choice of extra dig depth or extra power for digging in extremely hard ground. Due to the weight of the machine it was fitted with high-ply-rated 'balloon' tyres, which also provided maximum traction and flotation over rough ground.

JCB 4D

The 4D was launched in 1968 and was a larger, more powerful and more specialist machine than the 3D. Because the excavator was so large the 4D was fitted with a centrally mounted kingpost. The boom also had two different positions for mounting the dipper, one providing more speed and the other more power. The range of buckets offered was also unique to this model. It had only one stabiliser leg control

BELOW The 1967 JCB 3D was marketed as the excavator with a de luxe finish and saloon-car comfort. The cab was made of fibreglass and was said to be draught-proofed from floor level upwards.

lever, to reduce the number of levers needed by the operator, and featured a level link between the legs to keep them both at the same height. The operator was seated in an offset position at the excavator end to improve his view when digging. Parallel loader arms were offered as an option and the machine had a 4.8m (16ft) maximum dig depth.

The 4D had only limited success in the UK and most of the machines were exported to Australia, Canada, Israel and Scandinavia. Sales were fewer than anticipated – less than 200 were sold – so production ceased in April 1973.

Mk II range

The Mk II range was introduced in October 1968 and was identical in styling to the 3D and 4D that had been introduced shortly before. The new JCB 2, 2B, 3 and 3C machines retained their model name and were all similar in size to the models that they replaced. The Mk II machines were still painted in JCB yellow with a yellow cab, whilst the wheels, shovel and buckets were painted in Ayres red.

A new larger, squarer cab similar to the 3D was introduced in order to give more space. This incorporated an interior light as standard. The sealing rubber between the cab and the frame was improved to provide better weather protection and the cab's fixing bolts incorporated rubber bushings to reduce the noise level within the cab.

The excavator end underwent a number of

LEFT The 1968 JCB 4D was the largest backhoe loader JCB built, and due to its size it had a massive centre-mount kingpost plus extra stability from outriggers.

improvements, which included a reinforcing plate around the kingpost pivot and rerouting of the dipper and bucket hoses at the base of the boom to reduce flexing during digging.

New press-formed stabiliser legs removed the need for wear strips on the stabiliser legs, which were also thickened. Previously two 6V batteries had been fitted, but these were now replaced by a single 12V, 300A battery especially developed by Exide to JCB's own specification, in order to improve starting in cold temperatures.

As a protection against vandalism, hydraulic oil refilling was now secured under lock and key in a compartment within the offside fender. This ensured that all hydraulic fluid passed through the filter before entering the system, which prevented contaminated oil from getting in. The diesel filling point was moved from in front of the

BELOW The Mk II range of backhoe loaders was introduced in 1968 and featured a new, more spacious cab, improvements to the excavator end and full power steering.

29

DEVELOPMENT OF THE BACKHOE LOADER

RIGHT The 'High Lift' version of the 3C Mk II introduced in 1970 was able to load Atkinson road-gritting lorries.

windscreen to within the toolbox compartment in the nearside fender and a vandal-proof lock and key was added. Full power steering was introduced in place of the Hydrosteer power-assisted system, providing fingertip control.

A 'High Lift' version of the 3C was introduced in 1970, primarily for local authorities looking for a versatile backhoe loader capable of loading Atkinson road-gritting lorries.

Leyland 4-98 skid unit

The JCB 3C, 3D and 4D were built around BMC engines and the transmissions were made and assembled at Bathgate, Scotland. They were powered by a 4-cylinder 65hp diesel engine, whilst the 2B and 3 were built around the Leyland 3-cylinder, 45hp diesel engine.

Over a six-month period beginning in November 1971 further improvements were rolled out across JCB's entire range of backhoe loaders, which saw the introduction of the new Leyland 4-98NT engine to replace the 38TD model. This had improved combustion characteristics to reduce smoke emission and increase efficiency.

Mk III range

JCB's backhoe loaders remained almost unchanged thereafter until the introduction of the Mk III in February 1977, which saw the introduction of a number of improvements over the Mk II. Most significant of these were the changes made to the operator's environment. The Mk III was given a new cab with an opening lower half below the sliding door to allow walk-in cab access, which was a major improvement on climbing over the window frame. The cab was also 15cm (6in) higher, providing extra comfort and visibility and an improved driving position resulting from a raised steering wheel.

The main machine remained unchanged, however, except for increased slew performance from 127mm (5in) slew cylinders and better fill and dump characteristics from the new loader shovel profile.

Numerous further improvements made during production of the 3C Mk III included:

- Better engine protection and less maintenance courtesy of an improved air filter.
- Much-improved braking performance from 178mm x 102mm (7in x 4in) brakes.
- Maintenance-free, sealed for life front hub bearing.

BELOW A line-up of the JCB 1, 2, 3, 3C and 4C in 1964, showing a comparison of dig depths.

- High reliability sealed throttle cable.
- High reliability cold start and stop cables.
- Superior nine-speed synchro manual gearbox.
- Improved operator environment; cab soundproofing was fitted as standard, reducing cab noise by 5dBA (UK and Ireland).

JCB introduced its own transmission and axle in June 1979, which was referred to as the 'Powertrain' and 'Build 5'. It was used on the last 3Cs ahead of the introduction of the new 3CX. This was partly to try out the new transmission and get operators used to it, in readiness for the impending launch of the 3CX.

Model years		
	Production dates	
Model	First year	Last year
Trailer (screw type)	1945	1947
Trailer (hydraulic)	1948	1948
Major Loader	1949	1959
Si-draulic	1953	—
Hydraulic Trencher	1953	1954
Master Loader	1951	1957
JCB Load-Over	1953	1956
JCB Half-Tracks	1952	1957
Mk I Major Loader	1953	1956
Hydra-Digga Loader	1957	1960
Loadall 65	1956	1960
Loadall 75	1956	1960
JCB 4	1960	1962
JCB 4L (Loader)	1960	1962
JCB 4C	1962	1967
JCB 4D	1968	1973
JCB 1	1963	1971
JCB 2	1964	1968
JCB 2S (Shovel)	1964	1964
JCB 2B	1964	1968
JCB 2B Mk II	1968	1971
2D and 2DS	1971	1980
JCB 3	1961	1967
JCB 3 Mk II	1967	1977
JCB 3 Mk III	1971	1980
JCB S (Shovel)	1971	1980
JCB 3C	1963	1965
JCB 3CS (Shovel)	1963	1975
JCB 3C Mk II	1967	1977
JCB 3C Mk III	1977	1980
JCB 3C Mk III (Build 5)	1979	1980
JCB 3D	1967	1968
JCB 3D Mk II	1968	1977
JCB 3D Mk III	1977	1980

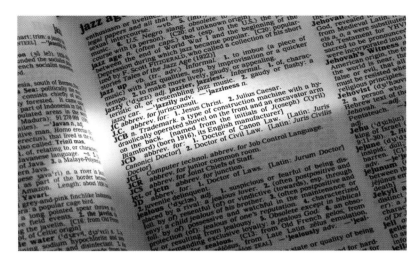

In the 19 years from October 1961, when the first JCB 3 was produced, until production of the JCB 3C III ceased in October 1980 (serial number 141351/5, sold to Algeria), nearly 72,000 backhoe loaders were sold in 135 countries. So phenomenal was its success that, in large areas of the world, the initials JCB have become the colloquial or generic name for all diggers.

The 3C Mk III had been JCB's best-selling backhoe model, selling almost 21,000 machines during its production years of 1977–80. However, it was becoming obvious that a replacement machine would soon be needed, and JCB had been secretly working on its successor – code-named Project 200 – since 1971. Production of all 3C III models ceased in October 1980, amidst a blaze of publicity as the new 3CX was introduced, starting at serial number 290001.

ABOVE The Collins Dictionary was the first to recognise JCB as part of the English Language, in 1987, and in 1989 JCB entered the Oxford English Dictionary.

LEFT The machines were ordered cab-less, and wooden boxes were fitted over the front controls to protect them. This is one of the very early drive-on vessels that transported large equipment as deck cargo. In the foreground a machine can just be seen coming up from the lower deck on a lift.

Chapter Three

Manufacture and sales

In 1954 JCB exported 17% of all of the machines that it manufactured, and by 1956 this figure had increased to over 25%. Sales in the UK between 1954 and 1959 virtually doubled year on year as JCB became a market leader in the manufacture of backhoe loaders. Despite a slight downturn in the market in 1961 due to a recession, sales bounced back in 1963 to almost 2,000 machines, of which over 30% were exported.

OPPOSITE A large order of several hundred 3Cs destined for Algeria leaving the factory on the start of their journey overseas.

By 1963 JCB had sold over 8,000 backhoe loaders and, with distribution in most European countries, overseas sales were becoming a major proportion of its sales and very important to future growth. The company's first overseas distribution conference was held at its Rocester headquarters in April 1962, at which orders for 413 machines were placed, worth over a million pounds. Exports continued to increase, and following a visit by JCB's Japanese dealer in early 1963 an order was placed to supply 300 backhoe loaders. This dealer had been distributing JCB backhoe loaders for two-and-a-half years and during this time had already sold over 100 machines.

By 1964 the backhoe loader concept had become well established and the UK market exceeded 4,000 machines, worth £11 million, which meant that JCB had earned itself a 59% stake of the market. Whitlock followed in second position with a 23% share, selling just over 900 machines, followed by International, Wain-Roy, Ford and Steelfab, who between them made up the rest of the market.

£250,000 staff bonus

The company had achieved a turnover of well over £12m, of which over one-third resulted from exports to practically every country in the world. Such was the growing success of the company that, in 1964, with sales up 60%, its employees shared in a £250,000 bonus. The news made national headlines and payouts were on such a scale that some employees were able to buy their first homes with the money that they received.

In the UK market JCB remained in pole position in 1967, but Massey Ferguson had climbed the ranks to achieve a 22% share and take second position from Whitlock, who by now had lost a significant share of sales and fallen to 10%. Ford had also now developed a foothold in the backhoe loader market, and whilst JCB maintained its number-one position with 56% of the market all of its other

BELOW The JCB 3C with its main competitors in 1971. From left to right are Massey Ferguson 2202, Ford 'Auto Dig', Whitlock 308 and Case.

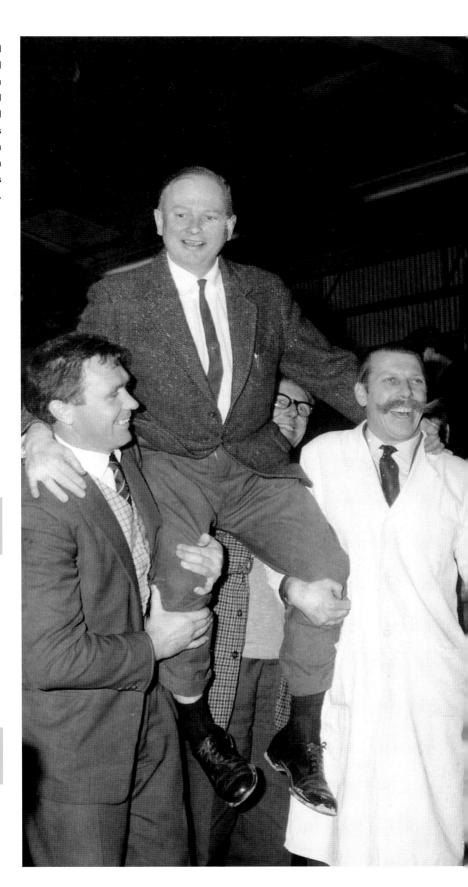

RIGHT In January 1964, JCB employees enjoyed their biggest ever payday as Mr JCB announced every worker would receive a bonus based on length of service with the firm. Payouts ranged from £40 to more than £500. The payouts had been made possible by the previous year's results, as sales had reached £8 million, an increase of 60% over 1962. Mr Bamford is seen here being carried on the shoulders of employees Bill Hirst MBE and John Wheeldon.

competitors had lost a proportion of their sales to Ford. Meanwhile the European market was growing at a rapid rate, with a market size by this time of 5,800 machines, of which JCB claimed 11.4%.

JCB's total sales of all backhoe loader models in 1967 was just over 3,000, of which 2,218 were sold in the UK and 800 were exported to Europe. By now the JCB 3C dominated the market, outselling the smaller, less powerful JCB 3 by 25 to one, while the larger 3D also increased in popularity.

By 1969 the UK market, in which JCB maintained its lead, had grown to 4,500 machines worth £17 million.

Queen's Award to Industry

In 1973 JCB still held first place in the UK backhoe loader market against its six competitors and was fourth in the world against a total of 30 competitors. More than 60% of its production was being exported to some 80 countries, so it's no surprise that in 1974 the company received the prestigious Queen's Award to Industry for increasing its sales from £11 million to £25.3 million by the end of 1973.

Market leader in 52 countries

Turnover by 1976 was £65 million and the output of each of its 1,500 employees was £44,000, four times the industrial average in Britain, over twice the American average and three times the European figure. JCB enjoyed

ABOVE 1972: JCB 3C Mk II at work in La Paz, Bolivia.

OPPOSITE At the end of the production line the machine would be filled with hydraulic oil and fuel before being driven off to have its shovel and bucket fitted.

Manufacturing efficiency

Despite fierce competition in the construction market, JCB's sales were strong. One of the main reasons that JCB remained very competitive was the high degree to which the company invested in manufacturing efficiency. In 1977 £5.1 million was invested in capital expenditure, and a further £6 million the following year. In 1977 sales topped £84 million, which was an increase of 29% on 1976, due partly to the range of heavy line products being produced (wheeled loaders, track excavators etc) and higher-specification machines.

World demand

JCB's backhoe loader sales had grown to 5,263 in 1977, of which a staggering 64% were exported. In 1979 the overall backhoe loader market was 3.1% larger than the previous year, with approximately 48,500 machines retailed by all manufacturers. Unsurprisingly the world market was dominated by the USA, which accounted for over 53% of sales, but the UK was the second-largest market with 4,032 machines sold in 1978, increasing to 4,508 in 1979 which represented almost 10% of the market. France followed in third position with 2,732 sales, equating to 6% of the total market.

60% of the backhoe loader market in the UK and was market leader in 52 countries throughout the world.

JCB's worldwide share of the backhoe loader market in 1979 was 12.5% compared with 11% the previous year, and its market share in the UK was 54.6%, up from 50.6% in 1978.

Key competitors at the time were Case, John Deere, Ford and Massey Ferguson, while Steelfab, Whitlock and Inchgreen had now ceased production of backhoe loaders. Although the top five manufacturers, including JCB, accounted for almost 89% of the world market in 1979, as many as 23 competitors made up the balance.

When production of the 3C Mk III ended in 1980 JCB remained market leader in over 50 countries. Over 72,000 JCB backhoe loaders had been sold worldwide and the company had developed a well-established dealer network. Taking advantage of its 28 years of backhoe loader manufacturing, 1980 was to prove a milestone in the company's history with the launch of a brand new model to replace the 3C Mk III – the 3CX. Laying the foundations for company growth in the 1980s, this represented a massive development and investment programme that had, cost in excess of £20 million.

For export regional differences see Appendix F.

Chapter Four

Anatomy of the JCB 3C Mk III

The most fundamental requirement of any earthmoving machine is a high-strength chassis capable of absorbing the stresses produced when the machine is working. This was the first feature that set the JCB 3C Mk III apart from its competitors. It enabled the engine and transmission unit to do their job without having to carry structural loads as well.

OPPOSITE In the foreground is the Leyland skid unit with torque convertor as fitted to the JCB. The lighter-duty skid unit of the Case backhoe is in the background.

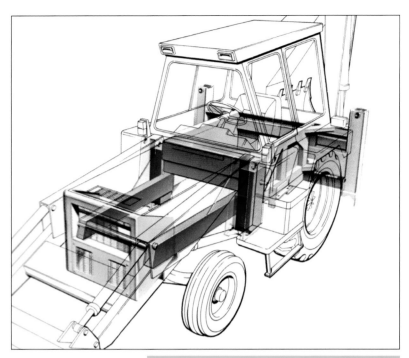

ABOVE The main chassis was called the 'Hydrachassis', as the hydraulic oil (blue) and diesel (green) tanks were integrated into it.

BELOW The 3C Mk III Hydrachassis.

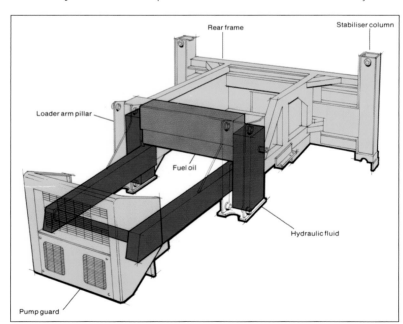

Hydrachassis

The 3C Mk III was built around the JCB Hydrachassis, which was an incredibly strong, rigid, one-piece integral frame constructed of all-welded thick steel box sections. These were stronger than the lightweight, pinned fabrications found on some competitors' machines.

The principle cross-bracing member at the top of the chassis was constructed mainly of 127mm x 64mm (5in x 2.5in) steel box section and was designed to incorporate the diesel fuel and hydraulic oil tanks. This therefore avoided the need for separate bulky and vulnerable tanks and coolers. The weight of the fuel and hydraulic oil was evenly distributed, providing a built-in ballast weight for the excavator and stability for the loader.

The system held 196 litres (43 gallons) of hydraulic oil, which allowed proper settlement time for separating and dispersing air bubbles, minimising the possibility of cavitations and thereby increasing pump life.

Chassis design

Shock loads from the excavator and loader were absorbed by the chassis and not passed on to other components, which therefore enjoyed much greater resistance to fatigue. This resulted in a much longer working life than conventional designs or tractor adaptations.

The chassis also provided a platform so that the hydraulics, transmission and other components could be neatly arranged for service accessibility and efficiency. It gave the machine a low profile for maximum machine stability and operator visibility.

Protection for the engine, radiator and main hydraulic pump was provided by a strong, fully welded guard at the front of the machine.

The one-piece chassis mainframe was designed to work with and integrate the following:

- Excavator frame with stabilisers.
- Loader arm and lift ram pivots.
- Mounting cradles for the engine and transmission unit.
- Front axle cradle.
- Box cross-sections of the chassis utilised for the fuel tank.
- Box cross-sections of the chassis utilised for the hydraulic oil reservoir.
- Lockable compartment for the hydraulic oil filter and filling point.
- Hinged step and lockable tool compartment.
- Cab mountings.
- Mud guards.

The engine and transmission unit were carried by the chassis, leaving it free from induced stress. Most competitors' machines used the engine/transmission unit as a chassis by adding a subframe, resulting in stress on the unit when the machine was digging and loading.

The inherent strength of the chassis enabled the machine to be jacked up off its wheels by the loader unit and stabilisers, so creating a firm working platform for excavating. In conjunction with the loading shovel, the hydraulic stabilisers were lowered to form a firm digging platform which absorbed the stress that would otherwise be taken by the wheels and tyres.

Excavator mounting

The rear frame, or 'H' Frame was of 152mm x 102mm (6in x 4in) box-section construction, incorporating vertical stabiliser legs into which the inner legs were mounted. The cast feet were made locally by Parker Foundry Ltd in nearby Derby.

Loader mounting

Pivot points for the loader arms and rams were located in places where they received maximum reinforcement from the chassis members. These immensely strong supports made it possible to use the loader unit for heavy-duty lifting and bulldozing.

Engine

The JCB 3C Mk III was fitted with the British Leyland 4-98 direct injection diesel industrial engine, producing 52kW (70hp) and maximum torque of 258Nm (190lb/ft) at 1,300rpm. The naturally aspirated version was first introduced in December 1972 on the 3C Mk II, commencing at machine serial number 64385, but had been used by BMC since 1965 as a replacement for the Fordson Super Major engine. It triggered off a development programme in which both companies closely co-operated to obtain a sturdy reliable engine and transmission. A Leyland transmission unit was also installed up until 1979, when it was replaced by the JCB transmission known as the Powertrain version.

Early engine problems

When first introduced to the earlier JCB 3 model the original BMC tractor engine was found to overheat in some cases, which could cause distortion of the cylinder head and block

LEFT 1979: This picture taken on the assembly track shows the chassis being lowered into position on the skid unit.

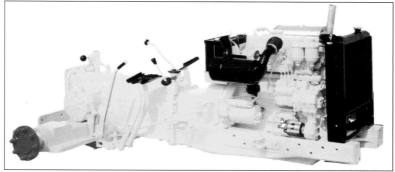

and a series of attendant faults. Liners were consequently thickened and modified. The block was stiffened to compensate for the reduction in 'bridge width' between the bores caused by the increased flange area of the liners.

The head and block were machined to finer tolerances and fitted with an improved cylinder head gasket having a steel backbone (introduced in June 1968) and extra reinforcing between cylinder bridges. Liners were eventually mounted in 'Siamese bores', ie pairings of cylinders 1/2 and 3/4. Studs were recessed and fitted with extra steel washers. Liners were subject to extra-critical inspection to ensure they stood proud of the surface, and rubber composite seals were incorporated. The water pump was also repositioned.

Despite the success of these measures, JCB wasn't satisfied that the problem had been

ABOVE Leyland skid unit with power-steering pump fitted on the side of the engine along with radiator. The ends of the rear axle were painted red to match the red wheels.

A cutaway illustration of the 3C Mk III, showing the internal components and driveline. This illustration was originally drawn by hand by one of JCB's skilled draughtsmen.

1 Bucket
2 Bucket teeth
3 Tipping link
4 Tipping lever
5 Bucket ram
6 Dipper
7 Dipper ram
8 King post carriage
9 Rear work lights

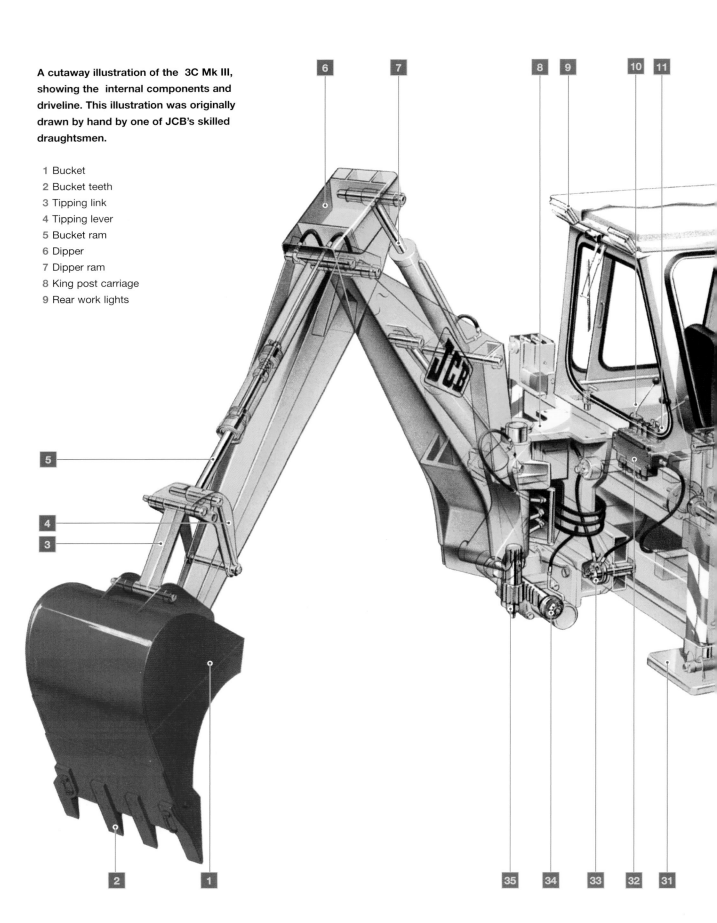

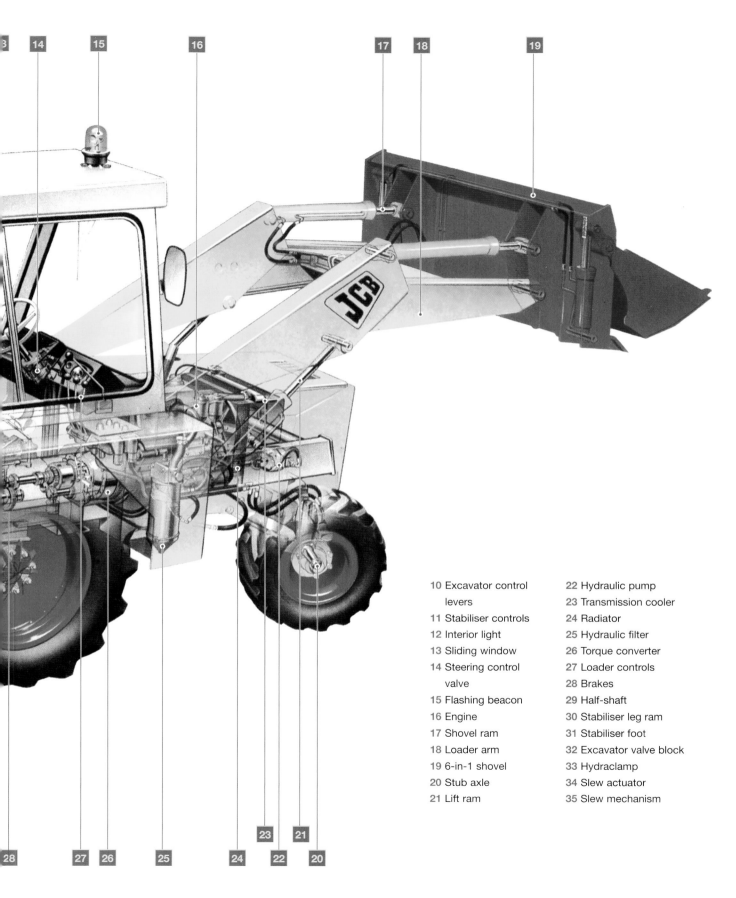

10 Excavator control levers
11 Stabiliser controls
12 Interior light
13 Sliding window
14 Steering control valve
15 Flashing beacon
16 Engine
17 Shovel ram
18 Loader arm
19 6-in-1 shovel
20 Stub axle
21 Lift ram
22 Hydraulic pump
23 Transmission cooler
24 Radiator
25 Hydraulic filter
26 Torque converter
27 Loader controls
28 Brakes
29 Half-shaft
30 Stabiliser leg ram
31 Stabiliser foot
32 Excavator valve block
33 Hydraclamp
34 Slew actuator
35 Slew mechanism

completely eliminated and so initiated a fresh development programme with BLMC (British Leyland Motor Corporation) that resulted in a completely redesigned engine with a four-port block and crossflow cooling system.

Introduction of the Leyland 4-98 engine

With the introduction of Leyland's 4-98 engine the following improvements were made:

- The section between the liners was increased by 45%.
- Additional liner seal ring was added.
- Stroke was increased from 120mm to 125mm.
- Stronger liner was used with increased wall thickness.
- Piston-head combustion chamber was redesigned to give reduced compression ratio and better combustion.
- Top piston ring was fitted with chrome insert.
- Exhaust valve area was increased by 19% and the valve made of higher-grade material.
- Exhaust valve seats were replaced.
- Injector sleeves were improved.
- Realigned cooling system by-pass hose was enlarged and realigned.
- New fan belt design.
- Sealed-for-life flywheel spigot bearing.

The benefits brought about by these changes included a torque increase of 6%, longer engine life, improved combustion, lower emissions, reduced maintenance and lower running temperature.

A four-plate cerametallic clutch was introduced for manual transmissions, which reduced wear rate, reduced slippage and improved cooling, resulting in longer life and better performance. A harmonic balancer and improved fan ensured smoother operation and substantially reduced oil temperature.

The engine's long stroke in relation to its bore provided increased torque, lugging power and flexibility over a wide speed range, all essential characteristics for heavy-duty earthmoving machines of this type.

Wet cylinder liners

Wet cylinder liners have most of their outside surface directly exposed to the engine coolant for maximum cooling efficiency. This overcomes the problems caused by different coefficients of expansion between the metals used for the liners and the block. The liners can therefore be made of extra-high-grade steel, enabling high piston speeds with minimum cylinder wear. Also, when bores eventually become worn replacing the liners is a relatively simple operation – a distinct advantage over reboring or replacement, which requires special equipment. Dry liners had the added disadvantage of needing to be of the same material as the block, with consequent increase in wear or restriction of stroke length, which resulted in inferior torque characteristics.

Cross-flow cooling

Water coolant was pumped to the main gallery

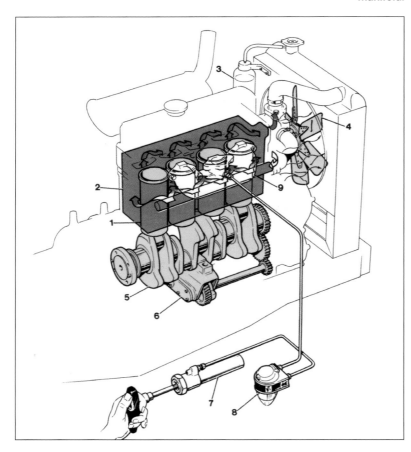

BELOW Four-cylinder Leyland 4-98 engine showing: (1) Wet liner; (2) Cylinder block; (3) Radiator expansion bottle; (4) Cooling fan; (5) Crankshaft; (6) Harmonic balancer; (7) Cold-start plunger; (8) Cold-start reservoir; (9) Inlet manifold.

and forcibly directed first to all four cylinder liners, where it was needed the most, before circulating over the head to the return outlet and back to the radiator. This eliminated the many problems inherent in the original engine's 'cold top, hot block' characteristics.

The system was a 'no-loss' type, whereby as the coolant expanded it overflowed into a container and returned when cooled. It comprised a plastic bottle mounted on the radiator, which eliminated coolant loss by evaporation and the need to top up.

Pusher fan

A six-blade pusher fan was fitted in order to drive heat away from the cab and prevent the dirt and dust caused in loader applications from being drawn inside. It was 406mm (16in) in diameter with 82mm (3¼in) wide blades.

Balancing

The engine incorporated a balancing unit driven from the timing gears. This counteracted the vibration caused by the one-way rotation of the crankshaft, which reduced noise for increased operator comfort as well as reducing vibration for smoother running and longer engine life.

Air cleaner

The engine was protected from dirt and dust by a large, dry-type air cleaner. The system featured a washable, replaceable paper element, a pre-cleaner for quick removal of collected dust, and a warning light on the instrument panel that signalled if the pre-cleaner became restricted.

Cold-start aid

An ether aerosol cold-start aid was fitted as standard. Combined with the heavy-duty starter, battery and high-charge alternator, this ensured easy starting in cold weather. To aid starting in arctic conditions a propane water-heater unit was offered as an option. This consisted of a thermostatically controlled heat exchanger and burner connected to the engine cooling system, a bottled propane gas supply and a timer.

The timer enabled the heater to be programmed to ignite at a pre-set time on a 24-hour clock, enabling the engine to be pre-heated before the machine was started.

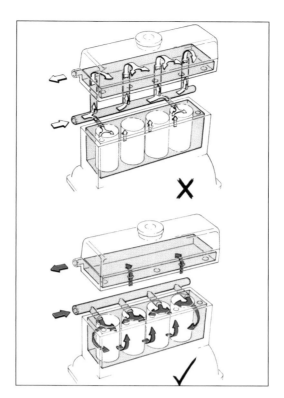

LEFT To prevent the engine from overheating, water is constantly pumped around the engine block and four cylinder liners where heat is generated from the combustion chamber.

Transmission and gearbox

Two transmission types were available: a standard manual gear change and clutch, and a more expensive optional torque convertor unit, which didn't require a clutch and was therefore easier to use.

The clutch was contained in a separate housing, which enabled access for servicing without having to split the engine from the transmission unit. The clutch could therefore be serviced in approximately one-fifth of the time it took on a unit without this facility.

The clutch was foot operated and incorporated a 330mm (13in) drive plate faced with cerametallic pads, which had superior wear characteristics and less slippage than the previous sintered bronze type. These pads formed a metal-to-metal contact with the pressure plate, providing exceptional performance and improved service life. Heat was efficiently dissipated through the channels of the segments and perforations in the drive plate, thus further contributing to increased clutch life.

The standard gearbox gave the operator a

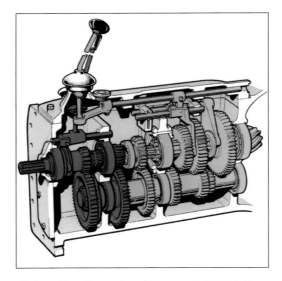

RIGHT The manual transmission comprised a simple robust mechanical gearbox, giving ten speeds on two levers.

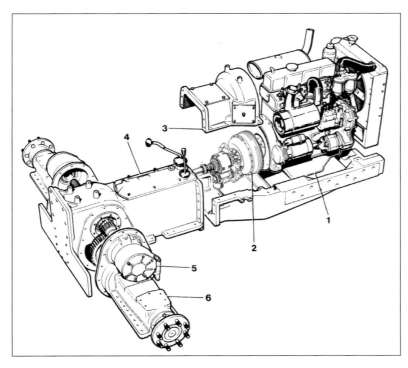

BELOW The BLMC (Leyland) industrial engine/transmission unit first introduced to JCB in 1965 comprised a manual gearbox and torque convertor. Key: (1) Engine; (2) Torque converter; (3) Torque converter housing; (4) Gearbox and final drive; (5) Brakes; (6) Rear axle.

choice of ten forward and two reverse gears to suit any job or ground conditions. It was a simple robust spur gearbox, which was very reliable.

A safety device, incorporating an isolator switch operated by the gear lever, prevented the engine from being started whilst the machine was in gear.

Torque convertor unit

JCB market research showed that, by 1975, 70% of annual backhoe loader production worldwide was fitted with torque convertors in preference to conventional clutch transmissions.

JCB's torque convertor was supplied by Brockhouse Transmission Ltd of West Bromwich, England. The 11N model was first introduced as an optional extra on the JCB 3C in June 1968, but the Lucas torque convertor was first offered as an option on the JCB 4 in 1961. It brought many advantages to the operator, particularly in terms of increased machine mobility, ease of control and smoothness of drive. It did, however, have its teething troubles, and required some development before it completely matched JCB's transmission and engine systems. Field problems were mainly attributed to occasional pump and oil-seal failures caused by insufficient filtration and convertor/engine misalignment.

The problems of filtration were overcome by the fitment of a suction strainer, a strainer in the filler pipe, and by drilling instead of coring the internal galleries. A flexi-drive plate and improved oil seal cured oil leakage.

3C Mk III torque convertor machines had one lever to select the speed range and another to select forward or reverse drive. There was no clutch pedal, which meant that conventional gear change and clutch work were eliminated.

The convertor was a fluid drive coupling in place of the clutch, which boosted initial engine torque by 2.8:1. Once a gear was selected, drive was automatic. Forward and reverse were obtained by simply moving the selector lever forwards or backwards.

The main advantages of the unit were increased daily output and reduced operating costs, achieved by:

- Increased engine torque when under load – *eg* driving away from standstill, climbing gradients and penetrating stockpiles.
- Exceptionally smooth take-up of drive, which in turn reduced tyre slip.
- Fewer gear changes, consequently less operator fatigue.
- Instant change from forward to reverse drive.

The gearbox was similar to a standard box but only used second, third, fourth and fifth gears. First, reverse and high/low were removed. The forward/reverse direction lever was adjacent to the steering wheel.

How transmission works

The term 'transmission' means transferring something from one place to another. In moving vehicles it means transferring the engine power to the drive wheels.

Nearly all machine vibration is caused by the engine or the usage of engine power. The 'link' in the transmission chain, which is vital to reducing vibration and shock, is the coupling between the engine flywheel and the gearbox.

Conventional clutches – however well designed – require moving parts to make contact with each other, the transfer of power being achieved through the friction between these parts. The only way to obtain a smooth drive is through the driver's skill in operating the clutch.

'Torque' means turning force. A torque converter uses fluid as the means of transferring power on the principle of one component driving another by the movement of oil. The components of a torque converter are three paddle-like discs – called the impeller, turbine and stator – working together within an oil-filled housing. A smooth, shockless drive is assured, and driver technique is less demanding.

The impeller is driven by the engine flywheel. Its vanes move the oil in the housing outwards on to the turbine. The turbine is rotated by the oil coming from the impeller. It provides drive for the road wheels through the gearbox. The stator comes into action when there's resistance to drive at the road wheels. In operation it redirects oil that hasn't been used by the turbine back to the impeller.

When the machine is in neutral or travelling steadily, the turbine is driven by the movement and force of the oil moved by the impeller. The rotational speed of the turbine is almost that of the impellor and nearly all the engine torque being developed is used. The oil itself in effect forms the coupling from the engine to the gearbox. The stator plays little part in this condition.

When there's resistance from the road wheels, the turbine is slowed down and rotates slower than the impeller. Not all the oil from the impellor can then be used and the pattern of oil movement changes. The turbine vanes are so shaped that oil that cannot be handled by the turbine is directed to the stator. The shape of the stator vanes enables this oil to be redirected to the impeller, and as the oil still has movement

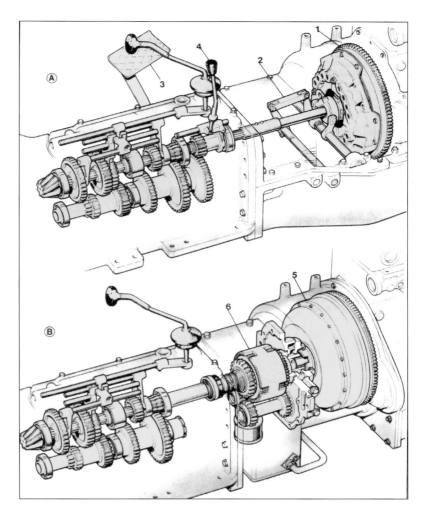

ABOVE Manual gearbox. A: Clutch and gearbox; (1) Dry clutch plate; (2) Clutch linkage; (3) Clutch pedal; (4) High/low selector lever. B: Torque converter unit and gearbox; (5) Torque converter unit; (6) Instant forward/reverse clutch pack; (7) Safety start – a safety device, incorporating an isolator switch operated by the gear lever, that prevented the engine from being started when the machine was left in gear.

BELOW Torque converter.

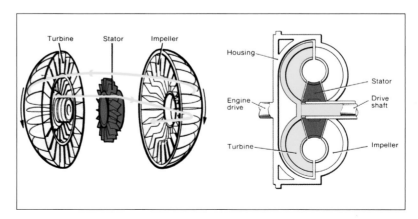

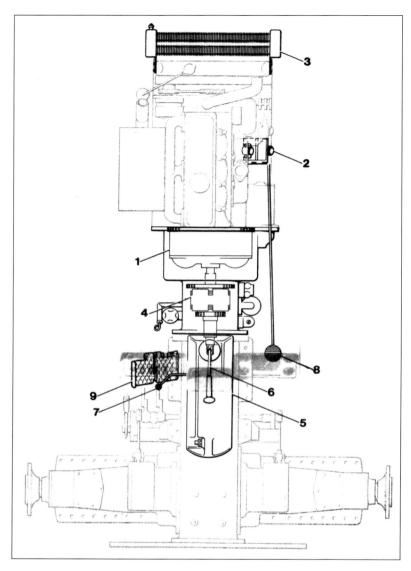

ABOVE Leyland skid unit fitted with Brockhouse torque converter. (1) Torque convertor: Located directly behind the engine, this comprised an oil-filled housing, turbine, reactor and impeller, which was connected to the flywheel. It provided a fluid coupling between the engine and gearbox. (2) Primary pump: The gear-type hydraulic pump was mounted in tandem with the power-steering pump and driven from the engine timing gear. It maintained oil circulation through the convertor for cooling and filtration purposes. It also provided oil flow for the operation of the clutch pack. (3) Oil cooler: A finned, tubed, air-circulation type oil cooler mounted in front of the radiator. This cooled the oil from the torque converter. (4) Forward/reverse clutch: Constant mesh gears incorporated twin, hydraulically operated graphite clutch packs. They changed the rotation direction of the transmission for forward or reverse drive. (5) Gearbox: The sliding-spur, four-speed gearbox provided four forward and four reverse gears. (6) Gear lever. (7) Direction lever: Located adjacent to the steering wheel housing, this actuated the hydraulic valve to direct oil to one or other of the twin multi-disc clutch packs. (8) Accelerator pedal. (9) Brake pedals.

it has energy. This energy is added to the impeller and gives it more torque to drive the turbine. Torque multiplication results.

Final drive

Rear axle drive from the differential to the road wheels was via a bull gear, therefore spreading the torque loadings and reducing component wear. The rear axle unit was of extremely strong construction throughout, ensuring complete reliability under demanding working loads. The differential gear had a built-in spiral bevel gear pinion drive with ratio of 4.38:1, and the final drive had a built-in spur gear with ratio of 4.182:1.

Brakes

The design of the rear axles enabled the brakes to be located on the high-speed (low torque) side of the final drive, where they were most effective. In this position they were also easily accessible for servicing, and as they were outboard-mounted they were protected from the weather and damage. The brakes on the later Powertrain version were inboard (refer to 'JCB Powertrain models' section on page 134).

Brakes fitted to Leyland skids were the dry, multi-disc type, and on the machines for most markets served as both operating and parking brakes. Additional drum brakes for parking were also available. The brakes could be independently operated, enabling braking on one rear wheel only, to assist in

BELOW The brakes fitted to the Leyland skid were dry and contained two brake discs on each side.

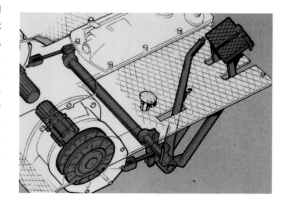

tight turning. This sort of braking was also referred to as 'steering brakes', and if the machine pulled to one side when using the front loader the operator could press on this brake to momentarily lock one wheel while the other continued to drive, therefore bringing the machine back to a straight line.

The outside diameter of the brakes was 165.1mm (6½in) and the total lining area was 1,216cm² (188.5in²).

Excavator and dipper

The boom and dipper were of strong, box-section construction, providing ample strength and rigidity to manage heavy payloads without undue stress. Both were heavy-duty welded fabrications and rugged steel castings, built-in at points where high stress would otherwise occur. These were, for the boom, the boom base, boom nose and boom ram eye end; and for the dipper, the dipper ram eye end.

The dipper featured a two-position bucket linkage offering a choice of either 'power' or 'speed' to suit the demands of the particular application. It was simple in design but of rugged construction, enabling it to withstand the extensive forces and loads to which it was subjected. It could dig at any angle within a 180° arc and remain perfectly stable.

Great emphasis was placed on 'fatigue life' design, which ensured that all the welds would sustain – without breaking – the varying stress concentrations to which they were subjected. After manufacture the boom and dipper were 'soaked' at 650°C in a large oven and then allowed to slowly cool down to relieve the stresses created during construction. This process, unique to JCB, doubled the life of the boom and dipper.

Performance was assured by creating narrow sections on both boom and dipper, which gave the operator good line of sight into the excavation and ensured maximum strength for minimum weight of metal. The main boom, dipper and bucket hoses were concealed with the boom and dipper along with the rams, to safeguard them from damage caused by striking the sides of the trench.

The excavator could be positioned anywhere across the rear frame to suit job requirements. This ability to position the arm to one side was

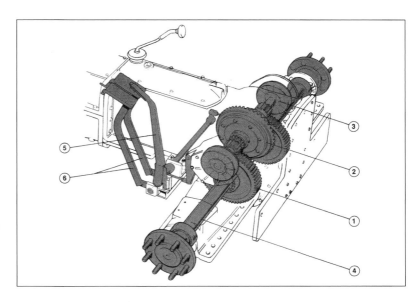

ABOVE Final drive. Key: (1) Bull gear; (2) Crownwheel; (3) Brakes; (4) Half shaft; (5) Handbrake; (6) Single/twin footbrakes.

BELOW Excavator. Key: (1) Rear frame; (2) Kingpost carriage; (3) Swing actuator; (4) Kingpost; (5) Boom; (6) Dipper; (7) Tipping lever; (8) Tipping link; (9) Excavator bucket.

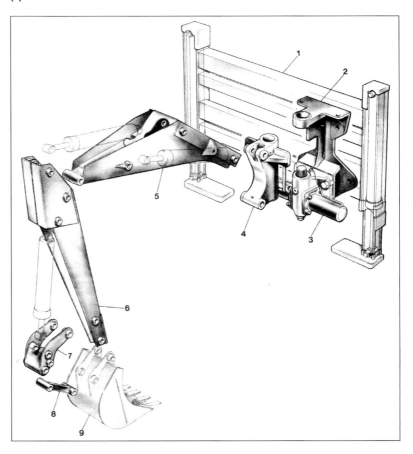

ABOVE The independent stabilisers allowed the machine to be levelled on uneven ground, to provide steady support during over-the-side loading. The underside of the machine was kept high off the ground to prevent damage when working on difficult ground.

particularly beneficial, since – as was mentioned earlier – it enabled building and trenching to be carried out in close proximity to walls and scaffolding. This feature was known as 'Hydraslide', a JCB-registered trade name. The Hydraslide design was first introduced on the JCB 3 in 1961 and was a key innovation, as some competitors at the time either restricted customers' choices of fixed peg points or didn't offer any at all.

Rear frame and stabilisers

The rear frame was the main support for the excavator unit, which was welded to the chassis. Two independent vertical stabilisers, mounted one to either end of the rear of the main frame, were hydraulically adjustable, enabling the machine to maintain full stability or to be levelled for excavating on slopes of up to 9°. The stabilisers also incorporated large cast steel self-aligning feet to spread the load over the ground and prevent them from sinking into soft earth.

Kingpost carriage

The kingpost carriage was the means by which the excavator was attached to and positioned on the rear frame, the kingpost itself being the mounting on which the excavator pivoted and slewed through 180°. It was a substantial iron casting made of spheroidal graphite iron, which was easy to machine accurately yet of great strength and toughness, sufficient to withstand the tremendous loads and forces to which it was subjected. The carriage was held in place by four nuts and bolts, one at each corner. These could be loosened either manually or hydraulically by means of the Hydraclamp (see page 51).

Extremely high quality bushings and washers were used throughout, *eg* a nickel bronze bushing 108mm (4¼in) diameter at the top, top and bottom thrust washers, and a nickel bronze bushing to take the fully floating boom pivot.

Slew system

The slew actuator which rotated the excavator on the kingpost was a rack and pinion type noted for its rapid but smooth, positive action. This enabled the operator to slew, dump and realign the bucket in a trench quickly and accurately. A rack and pinion system had been used on JCB backhoe loaders since the very first Mk I Major in 1953. On these early machines, however, the rack and pinion was 'open' and susceptible to wear from debris becoming trapped as it was positioned close to the ground. This was changed to a fully enclosed system in early 1961 on the JCB 3. Initially the slew actuators were positioned on the top of the kingpost carriage, but this was found to impede operator visibility to the rear and they were later moved to the bottom.

The slewing system gave constant torque through its full 180° of travel. Large-diameter pistons with end damping to cushion any tendency to over-travel were mounted on the kingpost to drive the rack against a pinion rigidly fixed in the kingpost carriage. The slew system was protected by two valves that controlled

RIGHT The kingpost is the mounting on which the excavator pivots and slews through 180°. The slew actuator, which rotates the excavator on its kingpost, uses a rack-and-pinion type design. This enables the operator to slew, dump and realign the bucket in the trench quickly and accurately.

acceleration and deceleration and acted as a system safety relief.

The oil-immersed unit comprised a rack with hydraulic cylinder at each end, which drove the pinion splined to the kingpost. The rack and pinion teeth were shot peened, and the whole unit was encased in precision hollow-extruded slew cylinders, giving full protection against weather and damage.

Hydraclamps

The Hydraclamp system was standard on the 3C Mk III and had been optional on earlier models that relied upon the operator manually slackening the nuts and retightening them. The Hydraclamp system simultaneously clamped all four points, which eliminated the risk of uneven or incomplete tightening. Hydraulic clamping of the carriage nuts also enabled the operator to control the entire Hydraslide operation without having to leave his cab.

The Hydraclamp system incorporated four expanding spacer clamps, one on each corner of the kingpost carrier, which enabled it to be clamped in position on the frame by hydraulic power. The four spacer clamps were connected in series by high-pressure hydraulic hoses, operated from the lever to the left of the excavator control valve block.

Hydraulic feed to the clamps was from the neutral circuit through the control lever on the excavator valve block. Pressure was maintained by a non-return valve. Moving the control lever to the 'off' position blocked the feed to the clamps and vented pressure in the circuit to the Hydrachassis.

Loader

The superior strength of the JCB loader, frame and mountings allowed for heavy-duty lifting, site-stripping and bulldozing, as well as providing a substantial working platform when the machine was jacked up in excavating position. The operator had good visibility of the front shovel, particularly in the 'carry' position when the shovel was at ground level.

The design and strength of the loader unit permitted the use of large hydraulic cylinders with a long stroke, providing both speed of working and power. Hydraulic pipes and

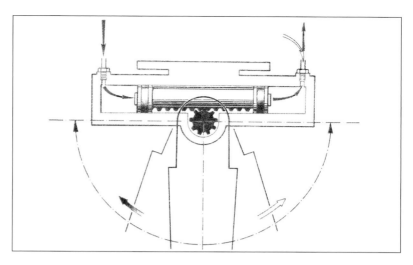

ABOVE Slew system. This oil-immersed unit comprised a rack with a hydraulic cylinder at each end, which drove the pinion splined to the kingpost.

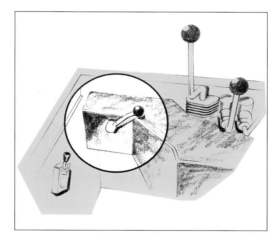

LEFT The Hydraclamp control lever is positioned on the left-hand side within easy reach and doesn't require the operator to leave his seat to operate it.

BELOW 3C Mk II fitted with a 6-in-1 shovel and industrial tyres.

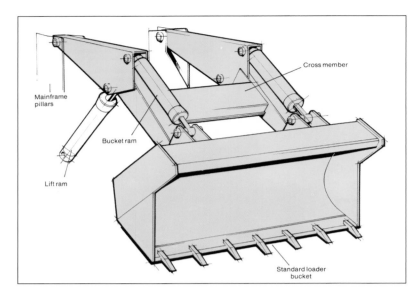

ABOVE A large crossmember braces the loader arms to prevent them from twisting.

BELOW The front axle was of high-quality material specification, with 57mm (2¼in) diameter kingpins in leaded phosphor bronze bearings. The heavy-duty forged stub axles operated in long-life, phosphor bronze bushes.

hoses were neatly arranged and well protected from damage.

Allowances were made in the system for the replacement of the standard control valve by a three-spool type, incorporating hydraulic levelling, enabling the fitment of other shovels or attachments that required an extra hydraulic service (such as the 6-in-1 clam shovel). Its main features were:

- Strength and rigidity to withstand the rigours of bulldozing and grading as well as normal loading.
- Hydraulic pipework protected from damage, provision also being made for additional service pipework for powering attachments.
- Standard pivots for interchangeability of JCB loader end attachments.
- Availability of a wide range of attachments including 6-in-1 clam shovel, jib crane and forklift.

The standard shovel was extremely versatile. It had a flat top parallel to the bottom edge so that the operator could see, at a glance, the angle of the shovel cutting edge in relation to the ground. The top edge was folded over, strengthening the shovel and providing a built-in spill guard to retain heaped loads at full height.

When used to lift the machine off its wheels into excavating position the shovel could be used in either level or rolled-forward position according to ground conditions. For example, the level position would be used if the machine was excavating from a road or level made-up surface, but would be inverted to give maximum stability on rough ground or when working from a hillside.

Like the excavator, the loader could accept many other attachments or optional equipment to attach to or replace the standard shovel. See Appendix C 'Attachments' for full details.

Front axle

The front steering axle was a substantial one-piece casting machined to accommodate the steering pivots and power-steering ram. Each steering pivot was an all-welded fabrication combining the wheel stub and track rod lever, machined to the correct steering angles and bearing tolerances. The axle also provided a proportion of the balancing necessary to counter the weight of the backhoe.

The front axle crossmember provided a strong mounting point about which the axle could pivot, which ensured that loads from the loader unit were taken up by the pivoted axle without any tendency to distort the chassis frame. The axle itself was a solid, one-piece casting having high resistance to such shock loadings as are inevitably encountered during heavy-duty loading applications. It had a static load capacity of 31,751kg (70,000lb) and dynamic rating of 10,886kg (24,000lb). Its enormous strength

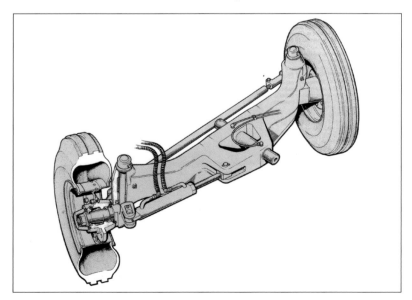

emphasised the vulnerability of the lightweight fabrications found on some competitors' machines at the time. The front axle also provided good wheel lock, which provided an unbraked turning circle of 7.6m (25ft 2in).

Axle-to-machine attachment was by a single pivot pin carried by steel pivot bearings, widely spaced to absorb torsional stresses, which allowed the axle to oscillate about the centre of the machine. It was pivoted in the axle yoke integral with the chassis, allowing 10° oscillation for stability over rough terrain. The long main pivot arrangement, which connected the axle to the chassis, provided exceptional fore-and-aft rigidity and high-load-bearing characteristics. Lubrication of the pivots, which in many vehicles are awkward to reach, was simplified by the provision of a remote greasing point on the machine's pump guard.

Positive steering under all conditions was ensured by mounting the steering ram directly on to the front of the axle, eliminating the need for ram oscillation between axle and chassis. This was an improvement from the earlier 3C Mk II model, where the steering ram was positioned along the chassis. Turn to lock was 2.75 turns clockwise and 2.15 turns anticlockwise.

The reliability afforded by this design was substantial due to its high-quality material specification. Large 57mm (2¼in) diameter kingpins were fitted into phosphor bronze bearings, each giving a bearing area of 1,806mm² (28in²). The bearings were also widely spaced to reduce loads. The heavy-duty forged stub axles operated in long-life phosphor bronze bushes.

LEFT A noticeable feature on the 3C Mk III is that its steering ram is fitted in front of the axle, unlike the Mk II where it runs in line with the chassis. The 50mm (2in) diameter chromed rod was deep induction-hardened to resist damage.

Wheels and tyres

With the 3C Mk III being used in so many countries and on so many types of terrain, it was necessary for a wide range of tyre alternatives to be available to suit all applications. The main considerations in selecting a tyre were the amount of road travel and on-site traction, the ground conditions, and the type of material being handled. Tyres are classified by type and size. The main types were truck, trailer, traction, industrial and ribbed. The ply was the only indication of a tyre's strength factor or carrying capacity.

Truck tyres were for the front steering wheels and continued road use. The cross-pattern tread provided good surface adhesion and an even wear life. They were popular on hard ground surfaces such as asphalt and concrete,

BELOW Tyre types.

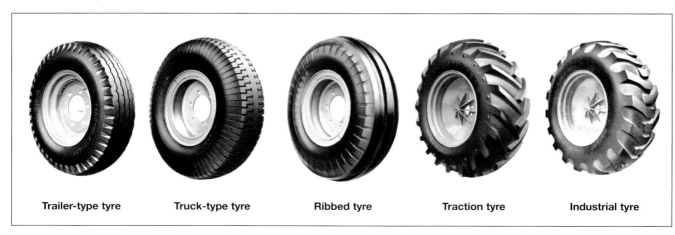

| Trailer-type tyre | Truck-type tyre | Ribbed tyre | Traction tyre | Industrial tyre |

which rendered them ideally suited to working in yards and on road duties.

Trailer tyres were for the front steering wheels and were intended mainly for continued road use, though they were also suitable for firm ground conditions. The longitudinal tread provided resistance to sideslip in deep ruts and on slopes.

Ribbed front tyres had deep, longitudinal ridges to provide positive steering on unpaved surfaces and slopes and over deep ruts. They were used in conjunction with traction-tread tyres on the rear driving axle.

Front traction tyres had narrower tread bars than the industrial type with more space between them, therefore attaining deeper penetration on soft ground. A softer rubber compound also made the tyre more flexible, which aided self-cleaning and was especially suitable for use on moderately soft ground. Rear traction tyres were designed for soft ground conditions and featured an open centre with angled bar tread, enabling a strong grip for high tractive effort and self-cleaning characteristics.

The industrial front tyre was an ideal all-purpose type with a large contact area giving low ground pressure. Earthmover tread gave positive grip on soft ground, good adhesion and long life in roadwork.

The industrial rear tyre was an all-purpose type for building-site preparation, roadwork and rough grassland work. It was particularly suitable for hot climates where ground conditions were hard. Interlocking lugs provided a wide area of contact and high resistance to punctures, guaranteeing long life on hard or abrasive surfaces.

Wheel rims, manufactured by Sankey of Shropshire, were classified by type, width and diameter. The type was indicated by letters conforming to an internationally recognised code.

Power steering

Most heavy earthmoving machines used some form of hydraulic power steering to reduce the effort required to turn the steering wheel, particularly when operating on harsh ground or carrying heavy loads. There were two main types: power-assisted, as fitted to the early JCB 3 models; and full power steering, as fitted to the JCB 3C Mk III.

Each type had its own hydraulic system comprising a pump, cylinder(s) and controls. The main difference between the two types was that with full power steering there was no direct mechanical linkage between the steer wheels and the steering wheel. The advantages of this were that the steering wheel was isolated from shocks at the steer wheels and there was no mechanical linkage to service. Steering also required less effort, providing finger-light, sensitive and positive control.

Power-assisted types retained the conventional steering column and direct linkage to the road wheels, which had the disadvantage that shocks from the road wheels were translated to the steering wheel. There were also more mechanical parts to be serviced.

Steering ram

The double-acting steering ram on the JCB 3C Mk III had a bore of 50.8mm (2in), a rod diameter of 25.4mm (1in) and a stroke of 196.8mm (7¾in).

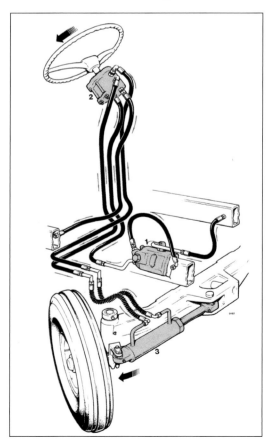

RIGHT Power-steering system.

Steering pump

The power-steering pump was a type C15 series gear pump manufactured by Plessey. It operated with a maximum system pressure of 120.8 bar (1,750psi).

Steering motor

The steering motor was an orbital steering Orbitrol OSPB 160 series type manufactured by Danfoss. The 'Orbit' valve, which was developed in 1958, gave vehicle producers more freedom of design due to the fact that it enabled the steering wheel to be mounted in the most convenient operating position, and the absence of linkage permitted ideal location of auxiliary equipment.

System operation

The pump delivered fluid to the Orbitrol which, if in the neutral position, allowed it to pass via the open centre servo valve to the return line. When the steering wheel was turned, the servo valve would direct the fluid to the appropriate port of the cylinder via the orbit gear set. This caused the inner rotor to rotate, thus metering the fluid and automatically restoring the servo valve to the neutral position.

Hydraulics

Right from the outset, JCB used fluid power to operate the loading and excavating functions of its machines. In the early days this approach was almost unique to JCB, but because the system had so many advantages its use soon became universal amongst earthmover manufacturers. Nevertheless, JCB used the knowledge gained in its pioneering approach to improve upon and extend the uses of hydraulic power.

The hydraulic system on the JCB 3C Mk III combined efficiency with simplicity. It was easy to control, enabling an inexperienced operator to perform efficiently in the shortest period of time.

Compared with most of its competitors, the JCB system was powered by a larger engine driving a larger hydraulic pump, giving a greater flow of oil to long-stroke cylinders, which added up to greater power and faster digging and loading speeds.

Components were of a high quality, being designed and built either by JCB itself or by reliable manufacturers to stringent JCB specifications. The patented two-lever excavator controls were a direct result of co-operation between JCB designers and Hamworthy Hydraulics Ltd of Poole in Dorset, and were first introduced on the JCB Hydra-Digga in 1959.

The hydraulics system was an integral part of the machine's design and the layout was tidy and functional. Sharp bends were avoided to prevent excessive strain on the hoses, which could have led to premature failure. Hydraulic pipes and hoses were fully protected within the body frame and the excavator and loader structures. The benefits of this were ease of maintenance, less downtime and less chance of damage.

The 3C Mk III had a basic hydraulic system, known as the neutral circuit, which included the fluid reservoir, pump and valve blocks, filters and their interconnecting pipes and hoses. When the engine was running, the pump drove the hydraulic fluid around this circuit without absorbing power from the engine, except that necessary to overcome friction. Excavator and loader end pipes and rams, and sometimes a third, auxiliary circuit, were added to the basic system. Operation of a control lever diverted fluid from the neutral circuit to a function circuit (*eg* the boom) and, because a ram offered resistance to flow, the pump absorbed power from the engine in order to overcome this.

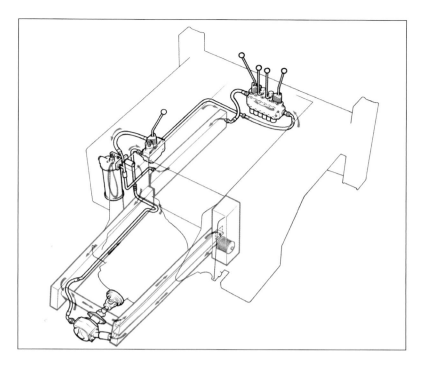

ABOVE Main hydraulic circuit.

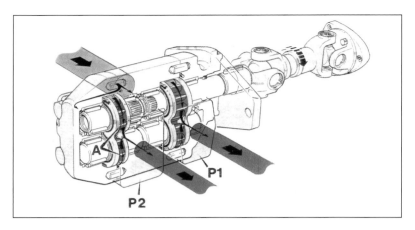

ABOVE The tandem gear pump contains two sets of gears that create two outlets for the hydraulic oil to flow, so that it can be directed to two different circuits.

The additional auxiliary circuit was fitted when a machine had the JCB Roadbreaker, Jaw Bucket or extending dipper option. This system had its own pump, which meant that the machine hydraulics were free to operate in the normal way, even when the additional equipment was in use.

System layout

Hydraulic oil was drawn from the Hydrachassis by an engine-driven gear pump. From there, it was pumped to the loader control valve block with was connected to the excavator control valve block by way of a high-pressure carryover pipe. The oil returned to the Hydrachassis through a full-flow high-capacity filter.

Hydrachassis

As described in the chassis section, the hydraulic oil reservoir was integral within the chassis members. The left-hand side member of the Hydrachassis was divided into two layers, which ensured maximum oil movement for maximum cooling, eliminating the need for a separate oil cooler. The total system contained 195.5 litres (52 gallons), of which the tanks held 118.2 litres (31.2 gallons).

Pump

The Hamworthy PA2210 gear-type pump was simple and reliable in design compared to other more complicated types. The gear pump was less susceptible to damage by dirt or water, easier to service and maintain and more economical when it came to replacing parts. The pump could be removed from the machine quickly, dismantled on the workbench for renewal of seals and wearplates and returned to service in minimum time. The hydraulic flow at system pressure was 126.1 litres per minute (33.3gal/min) and the system operating pressure was 138 bar (2,000psi).

The basic principle of the gear pump depended on the meshing of the two spur gear wheels, one of which was driven by the engine while the other was an idler. On the inlet side of the pump, oil was picked up and trapped in the space between the two teeth on both gear wheels and the pump body. As the gear wheels rotated this oil was carried around the pump until the space came into mesh with a tooth on the mating wheel. As the tooth filled the space, the oil was forced out of the space and through the pump outlet.

Loader valves

The 3C Mk III was fitted with either a single-lever two-spool loader valve, or a three-lever three-spool type. The single-lever valve permitted rapid operation for fast cycle times.

The two-spool loader valve block operated with a single lever only, which was required for lifting or lowering the loader arms, and rotating the loader arms and the shovel. This was fitted as standard to operate the general-purpose

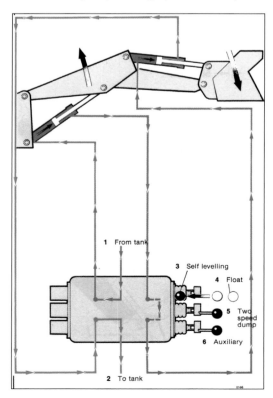

RIGHT The three-spool, three-lever valve block showing the path of oil through the valve block to the loader arm rams.

shovel. The valve block also incorporated the main relief valve (MRV) for the hydraulics.

The three-spool, three-lever valve block was fitted whenever an auxiliary hydraulic service was required, such as for operation of the clam action in the optional 6-in-1 clam shovel. It also gave the operator several other advantages over the two-spool valve, providing the following additional control facilities:

Shovel self-levelling
Hydraulic fluid displaced from the non-pressure side of a ram normally flowed directly back to the neutral circuit. Selection of self-levelling allowed some of the displaced fluid to flow to the head side of the shovel ram. This caused the shovel to extend gradually as the loader arms rose, so maintaining it at a constant angle relative to the horizontal. Self-levelling wasn't operational when lowering the loader.

'Float' position
The loader arms were allowed to remain lowered by their own weight, with the shovel resting on the ground. This feature was extremely useful for the cleaning up of undulating areas on completed worksites, as it allowed the shovel to follow the ground contour without cutting and disturbing the ground surface.

At the opposite end of the lift-spool travel was the float position. By connecting both sides of the lift rams to the neutral circuit and thereby creating a pressure balance, the loader arms tended to remain in their pre-set position, but would move upwards should the shovel strike an obstruction.

Fast dump
Shovel movement allowed fast, open dumping and accurate positional dumping. Fast dump functioned by using fluid from the rod side of the shovel rams to supplement that on the head side. This was done by moving the control lever beyond the normal shovel open position, which connected both sides of the rams to the high-pressure circuit. Because the working pressure was greatest on the head side (the largest surface area), the pistons moved outwards. Fluid on the 'rod' side was displaced towards the head side and the extra volume increased the rate of travel of the piston.

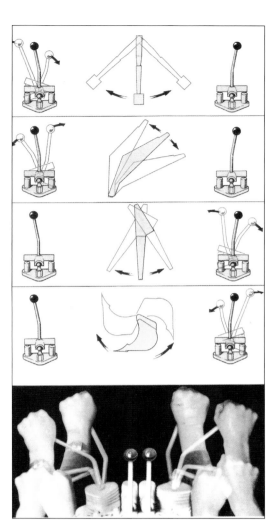

Extra hydraulic service
An additional circuit was provided for powering special hydraulically operated equipment, *eg* the 6-in-1 clam shovel.

Excavator valve block
The control levers mounted directly to the valve blocks, eliminating the need for extra linkages. This provided positive feel on the lever for the operator, eliminated the chance of delayed movement and reduced the need for lubrication or routine maintenance. The levers operated in a patented diagonal movement, giving simultaneous control of two services with one lever. ISO and SAE control patterns were available on export models.

Only two levers were required to operate the excavator. These were arranged so that one operated the boom and slew and the other the dipper and bucket services. The operator could

LEFT The two lever controls operate all the main excavator services without the operator needing to remove their hands from the controls.

ABOVE JCB invested heavily in machine tools and equipment that ensured greater efficiency of manufacture. This is the in-house chroming plant for the hydraulic rams.

high standards, and by 1976 JCB was producing, on average, 60,000 annually.

They were of thick wall tube with honed bores to ensure a smooth surface for the seals. The pistons incorporated a one-piece head, combining strength and eliminating possible leak paths. The chrome rods were JCB manufactured of tough, carbon manganese alloy steel and burnished to a fine finish prior to hard chrome plating to 0.03mm (0.001in) depth for maximum resistance against scoring and corrosion. The steering rams were induction hardened to a depth of 2mm (0.08in) for extra strength and wear resistance. Forged eye ends were CO_2 automatic welded to shafts for added strength and reliability.

The 'Selemaster'-type seals gave high efficiency over the wide temperature range of 40°C to 110°C. They had built-in flexibility to compensate for normal wear of the cylinder components and they expanded under pressure, thereby automatically adjusting the seal to match the pressure.

Rod end-damping

End-damping of the excavator bucket ram and dipper ram restricted oil flow at the end of the piston travel to automatically cushion any impact and protect the cylinder and associated components from shock loads that would shorten their working life.

As the cylinder neared the end of its stroke, a spring-loaded ring (1 in the accompanying illustration) on the rod side of the piston blocked off the main oil ports (2), leaving only two small apertures (3) for the remaining oil to escape. This had the effect of reducing the piston speed and cushioning the final movement.

therefore 'feather' the controls for balance, for a smooth movement of each pair of services. The control levers for the two stabiliser rams were also on this valve block between the two main levers.

Hydraulic cylinders

JCB manufactured its own cylinders, or rams as they're otherwise known, specifically designed for JCB machines. They were built to extremely

Hydraulic rams used on the 3C MK III						
	Bore		Stroke		Rod diameter	
	mm	in	mm	in	mm	in
Boom	114	4½	762	30	51	2
Dipper	102	4	737	29	51	2
Bucket	102	4	737	29	51	2
Stabiliser	92	3⅝	550	26	51	2
Slew	127	5	Rack and pinion			
Loader arm	102	4	762	30	51	2
Loader shovel	92	3⅝	594	23⅝	51	2
Steering	51	2	197	7¾	25	1

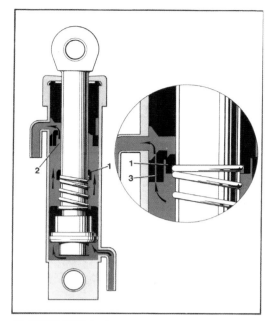

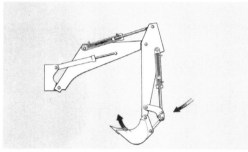

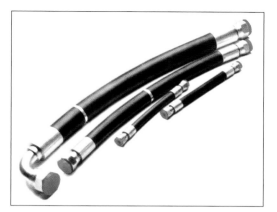

FAR LEFT Rod end damping.

LEFT JCB hoses went through the most stringent testing process to ensure their reliability.

LEFT Hydraulic oil fill is through the top of the main hydraulic filter. It was essential that the engine was switched off before filling otherwise oil would gush out of the filler cap.

Hoses

In the interests of performance and reliability, JCB fitted hoses of only the highest specification. Random testing was done on a purpose-designed test rig that simulated bending and pressure stresses up to ten times greater than those encountered under normal working conditions. This found that whereas a JCB hose resisted 540,990 impulses, the best of three competitors' hoses failed at 97,980.

Another feature of JCB hoses was that the number of sizes and lengths was kept to a minimum, ensuring the greatest possible interchangeability between machines. All hydraulic hoses fitted to the 3C Mk III were imperial sizes with BSP (British Standard Pipe) thread on the end fittings. Unsurprisingly there were many different types of end fitting to suit the position and angle in which the hoses needed to fit. Some had straight ends and others were at 90°.

Filtration

Earthmoving, as the name implies, is dirty work and dirt is the worst enemy of any hydraulic system. JCB therefore placed particular emphasis on oil filtration and introduced its own 'Filtromatic' system which played a vital part in maintaining the high efficiency and longevity of the whole hydraulic system. The basis of the JCB Filtromatic system was a dual-purpose filter located on the hydraulic return line at the filling point of the Hydrachassis.

This ensured that all oil was thoroughly filtered when the Hydrachassis was filled or topped up and continuously thereafter as it circulated around the system. The full-flow giant-size filter had an area of $3m^2$ ($32ft^2$) which filtered down to 5 microns. Each single filter could clean 5.2 million litres (1.2 million gallons) of oil between normal 400-hour element renewals, and the JCB filters were tested to five times the normal operating pressure.

Being located on the return line, the filter

eliminated any danger of oil starvation to the hydraulic pump. It was easily accessible for servicing, and by removing only one bolt the element could be quickly and easily extracted for cleaning or replacement.

The compartment for the filter was on the right-hand fender and was lockable for protection against vandalism. The filter was supplemented by a suction strainer in the Hydrachassis underneath, and the access plate was bolted in place near the step up to the cab.

JCB also paid careful attention to filter element materials and to the design of the container. Resin-impregnated paper – to resist attack by water absorbed by the fluid – was bonded by its ends and stiffening rings to make the whole assembly rigid enough to withstand five times the normal working pressure.

RIGHT The interior of the Powertrain cab is instantly recognisable, as the handbrake is positioned to the right of the front dashboard instead of on the floor.

BELOW Early 3C Mk IIIs featured a white steering wheel, which was later changed to black.

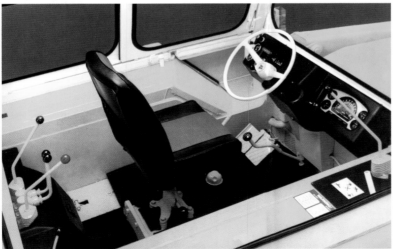

Cab

JCB recognised the importance of operator comfort at a very early stage, and pioneered the fitment of cabs to earthmoving machines as part of their basic design. Sales literature dating back to 1960 promoted the advantages of a comfortable cab that would make the operators, and therefore their machines, more productive. Early cabs were very primitive in design and did little more than provide a certain amount of protection against the weather, but as JCBs developed through the 1960s and 1970s more attention was given to cab design. Eventually operators were totally enclosed to protect them from inclement weather and cab heaters were introduced.

The whole cab area of the JCB 3C Mk III was an integral part of the machine, designed with the operator in mind. The ergonomic layout and controls enabled the operator to use the machine comfortably and easily, therefore enabling him to maintain a higher daily output with minimum fatigue. It had a large, uncluttered floor and the fenders provided useful flat storage areas at each side.

The clean lines of the JCB cab gave the machine a rugged appearance and the supporting framework was both strong and unobtrusive, ensuring that the operator had a safe, spacious workplace. Large areas of glass provided maximum visibility, and effective sealing ensured that the operator wasn't subjected to draughts or dust. Large, sliding windows were fitted to the 3C Mk III, giving good ventilation, and the entire rear screen of the cab could be moved from its normal position and stored against the right-hand-side window.

The cab framework was constructed of 1.2mm, 18SWG Zintec coated steel, offering protection against corrosion. The box-section pillars concealed wiring to the roof lights, windscreen wipers and piping to the windscreen washers where fitted.

For climates where a fully glazed cab wasn't necessary, machines were supplied with an open canopy that was bolted to the frame with ½in high-tension bolts.

White cabs

1972 saw a change from JCB's until then characteristic yellow cabs to white, which were phased in across the range of backhoe loaders over a six-month development programme. They featured a forward-sloping windscreen on the 3C and 3D due to frost not forming so readily on a downward-facing surface. They also provided more roof coverage for better operator protection. The overhang of the roof acted as a sun visor to protect the operator's eyes on bright days and, due to the angle of the screen, improved night vision too, as the dazzle from oncoming vehicles' lights was deflected away from the driver's eyes.

Caribbean cab

A great deal of emphasis was placed on further operator comfort in the form of the 'Caribbean cab', with reduced noise inside the cab achieved by the use of soundproofing material. The reduction of heat with the white cab and 'Sundym' green-tinted safety glass formed an effective barrier against solar heat transmission and offered reduced glare. The black-painted bonnet with matt finish minimised reflection that could cause discomfort and eye strain when driving on roads or operating the front loader. Sliding offside windows were added which supplemented the sliding door and rear window to give increased ventilation when required.

Safety cab

Legislation at the time didn't require safety cabs to be fitted except for certain markets. The special safety cabs that were fitted retained their clean, functional appearance by having the reinforcing members concealed within the existing framework.

Cab access

Access to the cab was through a sliding door on the left-hand side that was reached by two steps built into the mudguard, a stirrup-step on the fender and a large handrail conveniently placed on the cab pillar. The bottom, open-mesh step served as a mud scraper as the operator accessed the cab.

The door, which had a push-button handle incorporating a lock, was hung on a sliding track that had a smooth sliding action and was easy to open or close from outside or inside the cab. It could be safely left open to provide

ABOVE The 3C cab as delivered from the supplier ready for fitting the glass and trim. The cabs were manufactured by S. Whiteley and Sons Ltd of Cleckheaton, Yorkshire, who had supplied JCB's cabs since 1955.

LEFT The 'Caribbean' cab was developed for hot climates and was fitted with tinted glass to reduce the heat inside. The black bonnet minimised reflection when driving.

LEFT Open canopies were often chosen in hotter climates such as America and Africa, where the sun beating down on a fully glazed cab made it too hot for the operator to work inside.

RIGHT Access to the cab was made a lot easier on the Mk III, which had a low door that swung open to avoid the driver needing to step over the cab frame.

FAR RIGHT A large lockage toolbox was integrated into the side step for storing the tool kit and other essential equipment.

ventilation, and additional ventilation was provided by the rear window and the standard sliding window on the right of the operator.

Controls

The position of the loader and excavator controls was such that both could be reached simultaneously from the operator's central seating position. As well as providing immediate access in case of emergency, this also provided the skilled operator with accurate positioning over the point to be excavated, fast repositioning of the machine on long trench runs, and the facility to 'walk' the machine out from boggy ground.

On the JCB Powertrain the handbrake lever was moved from its normal floor position and mounted on the right-hand side of the control panel. A direct, centre-mounted gear change lever provided equal ease of operation, with synchromesh on all gears. Used as a clutch, the power disconnect pedal enabled effortless gear shifting.

Engine speed was adjusted by a cable-operated hand throttle mounted on the fender at the side of the seat.

Loader
The standard loader functions were controlled by a single lever within easy reach of the operator's right hand. This was cross-linked to a two-spool valve for all movements of the loader unit.

Excavator
Excavator and stabiliser levers were centrally grouped within easy reach of the seat. Operation was by the JCB two-lever system, each lever being connected to two spools in the six-spool excavator valve block by a patented cross-linking system. The left-hand lever operated the slewing and boom hydraulics, whilst the right-hand lever controlled dipper and bucket movement. Between the two were smaller levers that operated the stabiliser rams.

The small lever at the far left was used to

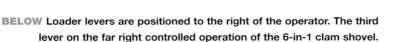

BELOW Loader levers are positioned to the right of the operator. The third lever on the far right controlled operation of the 6-in-1 clam shovel.

release Hydraclamp pressure when it was necessary to move the excavator unit across the rear frame.

Warning lights

Lights in the rear right-hand corner warned that the alternator wasn't charging or that there was insufficient engine oil pressure. The operator was therefore provided with all of the necessary features for the complete control of the machine from the rearward excavating position.

Seat

Several seat options were available, ranging from a basic type with height and reach adjustment to a de luxe full suspension type and, on some models, sideways movement.

The fully upholstered seat manufactured by Bostrom Saxon could be quickly adjusted for both forward and reverse facing. Height was adjusted by releasing the seat lock and rotating the seat either clockwise to increase the height or anticlockwise to reduce it.

Seats were mounted on a single pillar, which meant the changeover from loading to excavating controls was made by simply releasing a spring-loaded knob and swivelling through 180°. The location of the seat was high up in the machine centre, which provided the operator with an excellent all-round vantage point.

Suspension seat

An optional, semi-damped de luxe seat provided greater comfort. Manufactured by Bostrom Saxon, it was originally designed to improve comfort in unsprung vehicles, which normally transmitted all vibration and movement through to the operator. The full suspension seat, with its adjustable height, reach, back angle and support tension, absorbed vibration and shock loads and smoothed out the ride on rough ground and road travel alike. First introduced in 1971 as an option on the 3C Mk II range, it was fitted as standard to machines sold to the USA, Canada and Germany.

It was adjustable for reach, height and damping action and featured a hydraulic damping and suspension unit, which was incorporated into the backrest. The front edge of the seat was hinged and the rear edge

ABOVE The main controls for the excavator are positioned centrally. The throttle lever and Hydraclamp lever are within easy reach on the left and the pedals on the floor operate the jaw bucket or extending dipper where fitted.

LEFT The striking red and black seat was positioned centrally to give the operator excellent all-round visibility.

LEFT The optional de luxe full suspension seat provided far greater comfort, with adjustable height, reach, back angle and support.

ABOVE The JCB 3C Mk II had twin-lever controls for the excavator to reduce operator effort, and provided excellent all-round visibility.

BELOW The fact that the Massey Ferguson 50 backhoe loader was based on a tractor is evident in its cramped cab layout. The loader controls restricted access to the door.

was supported by a dampened suspension mechanism in the back panel. The seat framework was suspended at the back on an adjustable damping unit comprised of two springs and a hydraulic cylinder. A zip on the back cover gave access for adjusting the unit.

Certain installations also included sideways movement. As the seat was mounted on a single pillar, the changeover from loading to excavating controls was made by simply releasing the seat lock and swivelling through 180°.

Competitors' seat designs

Two types of seat were offered on competitors' machines. The first was a tip-over version where, to reach either the front or rear controls, the operator had to leave the seat to hinge it over. It lacked the comfort of the JCB seat and the adjustment was an awkward, laborious process, necessitating relocation of bolts in various holes or slots.

The second type had two seats, so that in order to change position the operator had to climb across the cab, which was tiring, time-consuming and inconvenient. The front seat was also generally low down in the cab, therefore preventing good all-round visibility. Where the separate excavator seat was raised, that part of the cab had to be raised too, resulting in an inconvenient double-decker layout within the cab.

Electrics

Alternator

Alternators were first introduced in 1972 on the 3C Mk II range to replace the more basic dynamo system, which did away with the need for a separate voltage regulator which became built in. The alternator produced alternating current, which was converted to direct current before being connected to the machine's electrical system. It was either a negative earth 12V with 43A-rated output or a sealed alternator option with 37A-rated output.

Lighting

All machines had brake lights, front and rear screen wipers and engine cold-start fitted as standard equipment. A cab interior light, two-speed heater and cigar lighter were also fitted for the benefit of the operator.

Road lights were to full automotive specification and met with the European 'E' regulations that came into effect from 1 January 1972, which required a change to the type of lens being used to meet the light emission standard. Both road and working lights were standard equipment on the 3C Mk III and included:

- Headlamps.
- Sidelights and directional indicators.
- Twin front and rear working lights.
- Licence plate light.
- Rear lights/reflectors/directional indicators.
- Internal light.

Instrumentation

Two different styles of front instrument panels were used on the 3C Mk III.

The one-piece rigid urethane instrument panel housed all the switches, gauges and warning lights necessary for the operation of the machine. The instruments were fully illuminated and easy reading. Two banks of fuses were located for easy access on top of the panel.

The standard instrument cluster included fuel level, engine oil pressure, coolant temperature and battery condition gauges surrounding a combined engine speed and hour meter counter. Torque convertor machines also had gauges for the torque convertor oil temperature and pressure, which were positioned on the left of the console.

Powertrain instrumentation

On Powertrain machines the control panel carried warning lights for all auxiliary functions except fuel level and engine hours. An audible and visual warning would appear should a failure occur, quickly alerting the operator so that appropriate action could be taken.

Optional and additional equipment

A number of options were offered as factory-fitted or aftermarket items that could be fitted by the dealer. Cab accessories included:

- Seat belt.
- Fire extinguisher.
- Vandal-proof instrument cover (first introduced on the 3C Mk II in 1975).
- Face-level fan.
- Steering-wheel knob.
- Coffee maker.

Rotating flashing beacon

A roof-mounted flashing beacon was available to warn traffic of potential hazards. It was weatherproofed to prevent corrosion of the internal parts and featured a rubber compound base into which the domed amber plastic lens fitted. A three-point bolt was fixed to the cab roof and a switch was fitted on the inside top left of the cab. The speed of flash was 90/100 per minute and it required a 12V, 36W bulb.

Hazard warning

A system was also available that enabled all four direction indicators to flash simultaneously, to warn approaching traffic of a potential hazard when the machine was working on public highways. It was operated by a single push-button rocker switch on the left-hand side of the front instrument panel.

The front and rear direction indicators were wired to a push-button unit mounted on the instrument panel. This button (push on, push off) operated all four indicators regardless of the position of the indicator selector lever.

Soundproofing

A sponge roof-lining and thick-gauge black rubber sheeting fitted on the floor and the

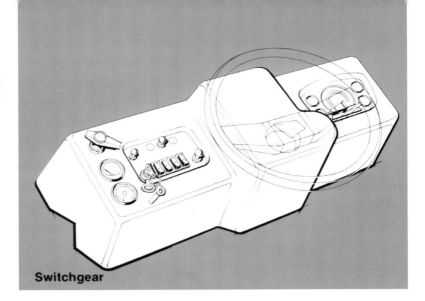
Switchgear

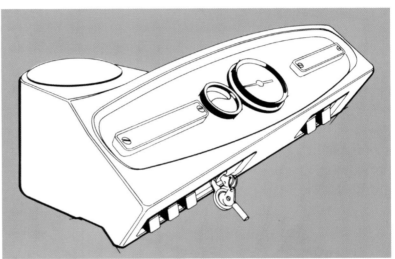

TOP The standard instrument control panel has a bank of rocker switches on the left-hand side.

ABOVE 3C Mk III Powertrain models feature a restyled control panel, and audible warning is wired into all monitoring systems to give the operator immediate indication of a fault.

RIGHT Bucket mounting brackets were available as an option for securing the buckets whilst they were carried in the front shovel.

BELOW A tarpaulin cover protected the cab. It is seen here fitted to a 1972 3C Mk II.

insides of the fenders came as standard, but an additional cab soundproofing kit was available as a factory option to externally reduce noise reaching the operator's ears from 95dBA to 90dBA. Due to the logarithmic scale used to calculate such things, this equated to a reduction of the noise level within the cab by 33% and meant less distraction for the operator and improved comfort. It was recommended when the operator was confined to the cab for long, continuous working periods.

There was also an engine soundproofing kit, consisting of a number of acoustically lined metal panels that simply bolted to the side and lower surfaces of the engine compartment. These panels, which could be easily removed for engine access, reduced noise emissions from the machine by 4dBA at 7m (23ft).

Headlamp guard
Headlamp guards were another option, to provide protection when the loader handled material that might fall on to the front of the machine.

Excavator bucket bracket
In June 1977 an excavator bucket mounting bracket was introduced which bolted on to the loader shovel using existing holes provided for this purpose. This bracket enabled the bucket to be carried safely in the loader shovel, where its extra weight over the front axle increased the stability of the machine when it was travelling on the highway or on site. The bracket holes were positioned so that it was possible to carry a 205-litre (45-gallon) oil drum in addition to the excavator bucket.

Exhaust extension
The standard exhaust only extended a short distance above the bonnet line, but an optional 40cm (16in) exhaust extension piece was available to direct emissions away from the cab. Fitting over the top of the standard pipe, it required no fixings to secure it. However, it needed to be removed before opening the bonnet.

Tooth guard
Required to meet road regulations in certain countries, this metal-formed guard provided protection from the shovel teeth and cutting edge. It was secured to each end of the edge of the shovel.

Heavy-duty battery
For cold climates, a heavy-duty battery was available at a 20h rating, 128 Ah, 520A.

Street pads
Introduced in August 1976, rubber stabiliser feet pads could be fitted to modified stabiliser feet and used when the machine was operating on easily damaged surfaces, such as soft tarmac and paving slabs. When the rubber pads became worn they could simply be unbolted and replaced.

Cab cover
A tarpaulin cover for the cab provided extra

protection against vandals, site damage and bad weather. When not in use it could be rolled up and stowed on the right-hand fender. The cover was supplied complete with padlock and necessary fixing strips.

JCB Roadbreakers

The Roadbreaker itself was a hydraulic hammer, which could be fitted with various cutting tools to suit the surface to be worked upon. These could be used to break up roads, concrete yards and footings, for rock-breaking in quarries, and for cutting and drilling applications.

They were first introduced in the UK on the 3C Mk II models in May 1974, which unfortunately coincided with a general mood of depression. Initial sales were consequently disappointing, but by 1975 had improved, with 93 machines sold fitted with a hammer kit. Almost three-quarters were exported overseas. Many customers were sceptical about the performance of hydraulic hammers and ignored them because of their lack of noise, mistakenly assuming that the hammer lacked power. As a result, it was found that the most effective selling technique was by a 'competitive' demonstration against a pneumatic breaker.

As part of the initial evaluation and sales period 24 Roadbreakers were fitted to machines sold to France, where the utility value of the equipment was appreciated, in addition to the savings in labour costs. Another 19 machines were despatched to the Middle East, where the hammers were used extensively for breaking up the hard ground prior to excavation.

Operation

Hydraulic power was provided by an independent pump driven by the engine operating at 103 bar (1,500psi). Two sizes of breaker were available, featuring double and rigid handles with a trigger on the underside of one of the handles that activated the breaker and then stopped it when the handle was released. Both were made by Montabert. The standard-duty BBH31 model weighed 31.75kg (70lb) and delivered 11.76kgf/m (86lbf/ft) power per blow, whilst the heavier BBH36 was 50mm (2in) higher, weighed 40.82kg (90lb) and delivered 13.83kgf/m (100lbf/ft). Both breakers used 22.75 litres per minute (5gal/min) and delivered up to 1,000 blows per minute. The noise level of the BBH31 model was 110dBA, compared to 109dBA on the BBH36.

ABOVE The hose reel for the Roadbreaker was introduced in October 1978 and could be fitted either side of the loader arms. The hydraulically operated Roadbreakers were more efficient than the pneumatic type and considerably quieter.

Advantage of hydraulic breakers

Hydraulic hammers weren't only more efficient than the pneumatic type (pneumatic systems wasted energy, as heat is created when the air's compressed), but they were also considerably

LEFT Twin hydraulic breakers could be used simultaneously whilst the machine was excavating to maximise productivity, and had many advantages over pneumatic hammers.

ABOVE Initially JCB was intending to use the American Digmore extending dipper, which was evaluated on the 3C Mk II model in 1973, but the company decided instead to design its own, better version.

temperature. It was also subject to controlled drying, which adjusted it for humidity before being circulated in the cab.

The 'Freon' or 'Arcton' gas-filled system provided 16,000BTU/hr at 80°F and 67% relative humidity. The current load had a maximum of 17A when the machine was stationary.

Cold start

For machines operating in arctic conditions where the standard ether cold-start system was insufficient, an optional propane gas heater system was available. This was coupled to the machine's water-cooling system. It pre-heated the coolant to assist starting at extreme cold temperatures and also provided quicker cab heating. The system was automatic, with a 24-hour timer that enabled pre-selection of the warm-up period (see illustration on page 69).

The unit principally comprised a heater, a gas container and a timer switch. The heater consisted of a heat exchanger and burner (4) and was connected by tubing to the engine cylinder block. The timer was wired to the heater via a solenoid valve (5) and was set to switch on one hour before the engine was to be started, then, at the set time, the timer closed the electrical circuit and operated the solenoid valve. This opened to allow gas to flow into the heater and at the same time passed current to the glow plug (6), which ignited the gas in the heater, which in turn heated the coolant and started the circulation.

When the temperature reached 130°C (266°F), the current to the glow plug was cut off by the exhaust thermostat and a holding current was supplied to the solenoid valve. A second thermostat in the lower part of the heat exchanger cut out the heater in the event of overheating.

The unit's propane consumption was low, at between 140gm to 200gm (5oz to 7oz) per hour and was therefore inexpensive to operate and, as it was non-electric, there was no drain on the battery.

Extending dipper

The extending dipper was a significant innovation on the backhoe loader and initially JCB was intending to use the American Digmore design, which was evaluated on the 3C Mk II model in 1973. However, by the end of 1974 the company had decided to move away from this idea and instead designed its own improved extending dipper.

First previewed at the River Authority Open Day in September 1975, the extending dipper option was introduced in the UK and Europe in April 1976 and for the export market in May 1977, adding greatly to the JCB's versatility and extra productivity by reducing the number of times the operator needed to move the machine during a working day.

The JCB extending dipper was a factory-fitted option that safely provided an extra

914mm (3ft) of reach, 955mm (3ft 2in) dig depth and load-over height, which permitted the use of higher-sided vehicles for extra load capacity as well as loading vehicles at a greater distance from the machine, which was useful when access was limited. It also saved time lost in manoeuvring over and between obstacles and in repositioning the machine and provided extra reach for river maintenance, site clearing and digging and levelling longer trenches from one position. Due to the strength and stability of the base machine it could handle a full load at full reach even to the side.

Operation took place from within the cab by means of two control pedals positioned alongside the operator's right foot; one pedal extended and the other retracted the dipper. The extension remained set when the foot was removed from the pedals and the extension was infinitely variable over its full length. The extension section was fully supported, even at maximum reach, and the hydraulic ram was positioned inside the dipper along with the hoses, which fitted with wire guards for full protection. To comply with tyre-loading regulations, 18.4/15-26 12-ply tyres were mandatory, as they had a higher load rating than the standard 28in wheels.

Hydraulic installation

When first introduced on the 3C Mk II models, the extending dipper circuit utilised hydraulic flow from the main gear pump. However, in September 1976 a new hydraulic circuit included an additional gear pump, mounted in tandem with the main system's pump and circuit to avoid power loss on other excavator functions. This meant that the unit could now be operated simultaneously with other services, ensuring fast and accurate control.

The necessary auxiliary hydraulic circuit was rationalised to suit Jaw Buckets and Roadbreakers as well. However, the Jaw Bucket couldn't be fitted to a machine with the extending dipper as it used the same hydraulic auxiliary circuit.

A double-acting hydraulic ram was used which took six seconds to extend and four-and-a-half seconds to retract. The ram bore was 69.8mm (2¾in), diameter was 38mm (1½in) and stroke was 1,067mm (42in).

ABOVE The extending dipper was introduced on the 3C Mk II in 1976 to dig and level longer trenches from one position.

LEFT The extended dipper was extended and retracted by foot pedals on the right of the cab, which enabled the simultaneous operation of the boom, dipper and bucket services.

Chapter Five

The designer's view

The JCB Hydrachassis proved to be a phenomenal success story, triggered by a major revolution in design. The Hydrachassis was pioneering in that it broke away from the traditional adapted tractor-chassis design, isolating the effects of the digging and loading stresses from the engine and transmission. It set standards of performance and reliability which have never been surpassed.

OPPOSITE A 1979 view of the design drawing office, which today houses the 'Story of JCB' museum. Ian McDonald (left) and Gordon Gray (right) can be seen in the foreground, with Tony Watson (white shirt, centre background) and Ian Stayner (striped shirt in background).

Torque and power

James Watt's original definition of one horsepower (hp) was 33,000ft/lb per minute. This tells us that one horsepower is needed to lift 33,000lb at a rate of one foot per minute. The work done in lifting this weight is a product of the force needed to raise it and the distance that the weight is lifted. The work done will be constant whatever time is taken to lift the weight.

Torque is expressed in Nm (Newton metres) or lb/ft (pounds force feet) and is the turning effort on the centre of a shaft, *eg* when a piston is powered down a cylinder it turns the crankshaft and produces torque. The greater the force (Newtons), on the piston, then the greater the torque. Also, the greater the distance (metres), between the centre of the crank and the big end, the greater the torque. Therefore torque = force x distance. In earthmoving equipment torque is usually more important than maximum power output. It's desirable to have a smooth build-up of power throughout the speed range and a good torque backup.

Efficient design

The JCB backhoe unit was developed over a number of years to make it the most efficient available in terms of both strength-to-weight and digging performance. The quality of its initial design, analysis and construction were essential to ensure long, trouble-free service. Great emphasis was placed on fatigue-life design, which ensured that all welds would sustain, without breaking, the varying stress concentrations placed upon them. The boom itself was 'soaked' at a high temperature to relieve stresses created during construction, and both boom and dipper were shot-blasted to eliminate surface strain caused during rolling of the original steel plate.

Steel castings were used at main pivot points to reduce stress concentrations and to distribute them evenly into the rest of the structure.

Pins and bushes

Pins and bushes were used extensively through the 3C Mk III and played a very important role in the working of the machine. Hollow-section metallic bushes were used with the majority of these pins, the pin rotating within the fixed bush to create a tight fit. It wouldn't have been sufficient to simply fit pins into the bored holes, because once the pin or member started to move considerable stress would have been placed axially and circumferentially, resulting in wear on the pin and, more seriously, on the walls of the bore, and when the pins or bore were worn they would become slack. Worn bushes, however, could simply be removed and replaced to restore a tight fit.

The bushes were of a different material to the pins, so that only one component wore and had to be replaced. The bushes were heat-treated to a lower hardness and were therefore softer, being deliberately designed to succumb to abrasion since the pin and shaft were much more expensive to replace.

Bushes were found in a number of places on the machine, including the front axle and the pivot pins connecting all of the hydraulic rams to the boom, dipper and loader. The pins were made of high-strength material to withstand breaking when subjected to forces such as bending. When initially put in, they were often knocked in with a hammer or a hydraulic press. It wasn't sufficient for a pin to be made from normal low-carbon steel, which couldn't survive dynamic and shear stresses or compressive and tensile loads. The excavator and loader pivot pins were therefore made from high-quality chrome molybdenum steel which was then radio frequency induction-hardened to a depth of 2mm (0.08in) to increase its strength, reduce wear and withstand the extreme loads that would act upon it. The induction hardening gave a hard outer skin to resist wear, while retaining a tough core to resist impact loads.

RIGHT The pivot connecting the boom ram to the boom had a cast pivot boss welded to the outside to provide additional strength. It was shaped like a keyhole to avoid a concentration of stresses that could have led to fatigue failure.

Tension bushes

The majority of the pivot points had a greasing point to reduce friction and wear between the pin and bush. The bushes fitted to the bucket pins were special CV tension bushes, which were made from special spring steel and had a longitudinal cut to keep the bush tensioned outwards on the walls of the bored hole, so preventing the bush from turning in the bore and causing wear. A flange on one end of the bush prevented it from moving inwards. The opposite tapered end made it easier for the bush to enter the bore. Where two were placed together, taper to taper, they formed a cavity for grease to enter.

Grease reduced friction to a minimum, but did have the disadvantage of attracting dust and foreign matter, which could be drawn into the bush and pin workings and cause wear. Nevertheless, grease created a protective film over the part, and pumping new grease into the bush pushed out the old contaminated grease carrying foreign matter.

Performance

Performance was assured by creating narrow sections for the boom and dipper, which gave the operator good line of sight into the excavation and offered maximum strength for minimum weight of metal. Concealed rams and hoses were also protected against damage caused by striking the side of the trench.

Robust links and levers combined with powerful ram effort to produce high breakout force at the bucket toeplate. The tipping link could be supplied with a lifting shackle for lifting applications.

The excavator and stabilisers were mounted on an integral frame, which, in turn, mounted directly on to the chassis, providing great strength and rigidity that absorbed the working loads and allowed a clean, fundamental layout. Combining the main chassis and the rear frame into a fully welded unit ensured that the loads generated by digging and loading were accommodated and were effectively distributed through the substantial chassis and solid cast front axle, relieving other components from stress. To ensure that this happened without creating unacceptably high stress levels, the

ABOVE The narrow boom and dipper design of the JCB enabled excellent operator visibility of the trench.

BELOW During manufacture of the chassis the rear stabilisers and Hydraslide are held in a large manipulator that can be rotated to enable welding all around.

ABOVE A 3C Mk II chassis in the yard, ready to be taken to the production line.

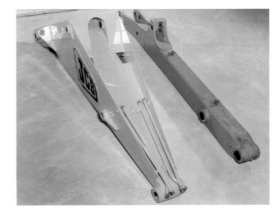

RIGHT The size and inherent strength of the JCB 3C boom (left) can be seen clearly when placed alongside the lighter-duty boom of the Case backhoe loader.

structure had to be carefully analysed at the design stage. This meant that, long before a machine went into production, it had already been thoroughly tested to make sure that it could survive the rigours of its intended use.

Excavator design

The JCB excavator arm, or backhoe, was developed over a number of years to make it the most efficient available in terms of both strength-to-weight ratio and digging performance. The basic design of the JCB backhoe mainframe proved itself over many thousands of machines and remained fundamentally unchanged. It continued with the twin beam arrangement, cross-braced by the rear frame, fuel tank and axle trunnion bearer. To allow an increase in steering angle, the side members carrying the hydraulic tank and the engine front mounts were angled inwards.

The structures were designed on the 'fatigue-life' principle, which imposed much more severe restrictions than straightforward static stress considerations. The boom and dipper were subjected to the full digging stresses of a machine designed for high performance and were therefore built to absorb great applied hydraulic force and the weight of heavy loads lifted at full reach to full height.

The main members were of 10mm (0.39in) or 12mm (0.47in) plate, with steel castings at high load points, stress-relieved after welding – a heat process that eliminated inherent metal stress built up by welding. Well-supported pivots, with no pins less than 44mm (1¾in) diameter, added to overall strength and resistance to wear. Pivot pins throughout were electrically heat-treated to a high wear resistant, non-brittle finish. This process, known as induction hardening, hardened the outer casing of the pins to a predetermined depth.

Boom and dipper design

The machine's tremendous digging power subjected the boom to considerable shock and impact loads. Stress-relieving heat treatment was used to eliminate the stresses set up in the components by extremely high local temperatures attained during welding. Stress-relieving entails heating the components to a temperature of 650°C (1,202°F) very gradually over a period of 16 hours, to ensure that the whole thickness of the metal is evenly heated. After being kept at this temperature for one hour, known as the 'soaking period', the oven is allowed to cool very gradually. This treatment enabled components to absorb shock loads safely, tests establishing that the working life of such parts was increased by 100% to 150%. The boom, dipper, tipping lever and link, kingpost and carriage were all stress-relieved to improve the life of their components.

The boom and dipper were made as wide as possible at the pivot points, which provided a firm anchorage to counteract any tendency to twist when under load. The boom incorporated substantial castings at the base, dipper ram pivot and nose, and pivoted to the kingpost in a nickel bronze bush via a fully floating pin, free to rotate in the bush and thus prevent uneven wear. The dipper pivoted in a phosphor bronze bush in the boom nose casting. Its ram incorporated a feature called rod end damping – a device that rapidly slowed down the speed of the ram as it neared the end of its stroke, thus avoiding damage due to careless use of the controls. The dipper carried the bucket lever and link. All three pivot points of the lever were fully floating.

The bucket featured two link positions on the bucket lever providing maximum tear-out force and bucket rotation. In the power position, the lever exerted maximum turning force for full bucket penetration power when working with extra hard material. In the speed position it gave the bucket a rotating arc of 201°, offering two major benefits: the bucket wrist action could be used to give a clean square cut at the bottom of the hole, almost to the full extent of the excavator's digging depth; and, when raised to full height, the bucket could be fully angled to prevent spillage of load.

The JCB could therefore handle heaped loads, giving sustained high production with less cleaning up. The bucket itself had a specially curved shape that gave a radial action, concentrating full power at the teeth for maximum penetration.

The diagram shows how the simple principle of leverage was applied to utilise the work of a cylinder to a different effect (work = force x distance). By altering the leverage point the amount of work could be utilised to provide either a large force over a short distance, or less force over a greater distance.

Loader design

Simplicity and strength were two features carried through the whole design of the 3C Mk III and were most apparent in the loader unit.

The arms were fabricated from 5mm ($^3/_{16}$in) thick, U-shaped steel, forming a box by the addition of a fully welded bottom plate. Each structural component of the loader arm was of triangular profile for maximum rigidity. The side arms were braced at the outer end (nearest the shovel) by a deep, webbed single box section crossmember called the torsion box, which made the loader arms resist any twisting.

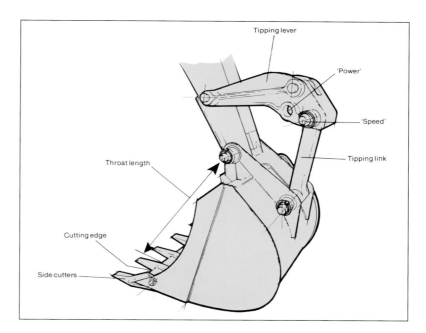

ABOVE Unique to JCB, the tipping link had two pin positions to maximise both bucket rotation and tear-out force.

LEFT 3C Mk II working in the Caribbean in the early 1970s. The operator is standing in the foot-well to give him better visibility of the deep trench being dug.

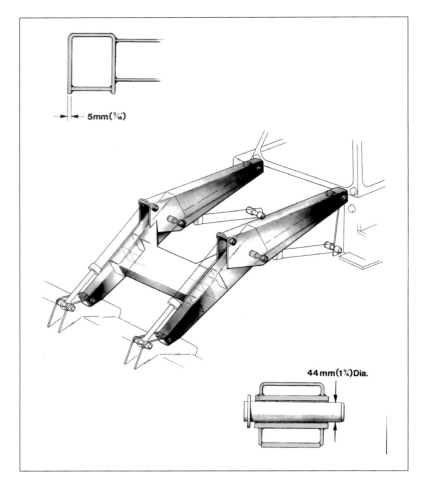

The simple design geometry enabled a reduction in the number of pivots compared with a more conventional system, the JCB loader having just 12 pivot points – less than half the number of some of its competitors – which consequently reduced both servicing time and cost.

Pivot pins for the arms, cylinders and bucket linkage were 44mm (1¾in) diameter steel, EN19C induction-hardened for maximum wear resistance and supported on each side by heavy-duty steel bosses.

Testing the product

In the late 1950s J.C. Bamford's policy was to carry out more laboratory testing, which could be carried out at high speed so that a component's full working life-cycle was completed in a few weeks rather than the several years it would have taken working on a site. After a careful search, orders were placed for the German Schenck Servo-Hydraulic Structural Testing Equipment, capable of subjecting the main chassis to stresses that would be imposed on it in service but at much higher speeds.

JCB set up a separate company, JCB

ABOVE Simplicity and strength, two features carried through the whole design of the JCB backhoe loader, are most apparent in the loader unit.

RIGHT Testing of new products (this photograph was taken in 1962) would be carried out against competitors' machines to benchmark the product and determine its advantages, which the salesmen could then explain to prospective JCB customers.

Research, to deal with all matters of research and development. This comprised a highly qualified, carefully picked team of specialist engineers, designers, draughtsmen and technicians, who were engaged in constant research to improve and extend JCB's range of products. Each proposed machine was fully discussed at the highest level amongst the company's financial analysts, estimators and other experts to balance both engineering and commercial requirements.

The ideas were next converted into drawings, then to models, mock-ups and prototypes which were rigorously tested in JCB's well-equipped experimental department and in the field before being passed as fit for production.

Structural tests facility

December 1976 saw a significant step forward, with the investment of half-a-million pounds into completing phase two of JCB Research's structural test facility.

By 1979 JCB spent, on average, 4% of its annual sales revenue on research and development facilities, compared with an average of less than 1% for the UK construction equipment industry as a whole. The multi-million pound test centre lacked for nothing in the way of modern, scientific apparatus and equipment for investigating and evaluating the performance of components and assemblies.

Destruction test rig

The destruction test rig was designed to carry out accelerated life tests on many of the components used in the manufacture of JCB hydraulic machinery. A complete machine was mounted in the rig and set up in the normal digging position with all wheels clear of the ground, thus leaving the entire load supported by the excavator rear frame and the front loader shovel.

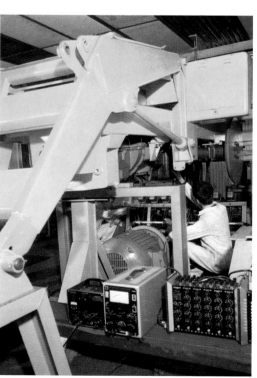

ABOVE Models of all concept machines were made out of wood. Here Mr Bamford is seen discussing the JCB 3C in the model-making room.

LEFT 3C being subjected to concentrated testing on the in-house test rig, which worked 24 hours a day.

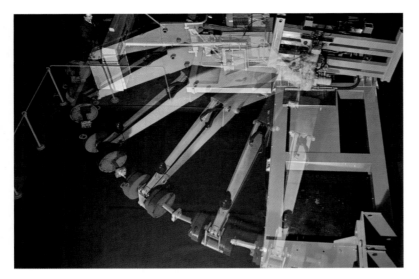

ABOVE **Slew test rig.**

The excavator was set up with the boom, dipper arm, link, lever and bucket positioned in such a way that all the hydraulic rams were acting at their maximum leverage. The excavator bucket was securely fastened to the base of the rig so that, during the test, the whole structure was held fast. The only movement would've been that of the flexing of components under stress.

BELOW **The very first hose testing rig in 1961, showing Mr J. Leake, a development engineer, taking a pressure reading. The rig was designed to subject the hoses to extremes of pressure and flexing for prolonged periods.**

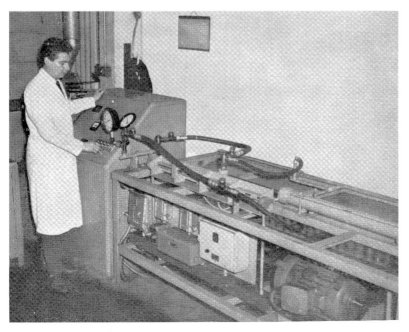

Oil was fed from a pump driven by an electric motor at maximum operating pressure. The high-pressure oil was fed into a bank of solenoid-operated control valves, which fed the oil to each of the four hydraulic rams in turn in the same sequence as for a normal digging cycle.

The solenoid valves were electrically operated by a series of micro switches that were operated by cams. The cam could be so arranged as to give any desired sequence of digging operations. For instance, it's known that in practice many high pressures could be induced in certain rams by the operation of other rams, particularly when the boom, dipper arm and bucket are in a particular configuration. These conditions could be reproduced on the test rig by inserting additional cams to produce induced pressure as and when required.

Hydraulic test rig

As hydraulics played such a vital function in the design of JCB machinery, it was natural that research into the application of hydraulics formed a major part of the work of JCB Research.

The hydraulic hose fitted to any item of earthmoving plant is of vital importance. Continual hose failure results not only in costly downtime, but very often in other plant also being brought to a standstill. The benefits of an efficient and reliable hose were therefore obvious. Consequently JCB had several permanent hose test rigs, and as hose improvement was a continuous process new hoses were always under test. The results obtained from these rigs gave details of length of life, effect of flexing and effect of pressure waves. Occasional straight burst tests were carried out, but these were only to maintain a check on the four-to-one minimum safety factor.

Ram test rig

To test hydraulic rams, two rams were used – one under test and the other acting as a slave to provide a working resistance. The two rams were on fixed pivots at one end and both piston rods were mounted to a common central pendulum. Oil flow from the slave ram was restricted to provide resistance for the test ram.

The stroke of the unit was controlled by micro switches that operated solenoid valves, which in turn pressured the piston.

Test cells

Special test cells were installed that simulated wide extremes of environmental conditions individually reproducing equivalent altitude conditions up to 3,000m (10,000ft) above sea level; sub-Arctic temperatures as low as -40°C and tropical temperatures as high as 52°C. By adding sound-absorbent foam linings the cells become anechoic chambers for accurate noise evaluation.

Quality isn't something you can add to a product as an afterthought. It has to be planned at the design stage, built in to the manufacturing process, and monitored continuously. So it was with JCB, from raw materials to final assembly. Component inspection was unusually thorough, and was 100% on many items, especially on high-workload items such as gears, hydraulics and brakes.

Part numbers

Due to the range of machines being designed, an extensive quantity of part numbers was being introduced on a daily basis by the engineering drawing office, and a new part-numbering index became necessary. A revised group prefix part-numbering system was introduced in 1964 in which the first four digits identified the part type. For example, sales publications and promotional items were allocated the prefix 9999/ and service publications 9801–9802. Every part on the machine was identified in the same manner; for instance, the standard parts and hardware designated for pivot bosses is 1050/–1154/ and metric nuts are 1370/. All the hydraulics hoses have the group prefix 610/ to 639/, and the following hose prefixes are used on the 3C Mk III:

611/	¼in BSP
612/	⅜in BSP
613/	½in BSP
614/	⅝in BSP
615/	¾in BSP

ABOVE The 3C Mk III undergoing extreme cold-start engine testing in JCB's own cold cell testing centre at the factory during 1977.

BELOW All machines went through a factory hot test and pre-delivery inspection (PDI) before they left the factory to go to the dealers.

Chapter Six

The owner's view

If you ever think of buying a 3C Mk III to restore or to use as a working machine, there are several considerations to be taken into account before you make your purchase. If your intention is to use the machine for building a house or clearing a piece of land or such like, then first assess how much work is required and take this into consideration when viewing the machine so as to ensure it looks up to the job. If it's your intention to drive the machine on the road between jobs, then a logbook will be required to register it for road use, and the machine will need to be in an appropriate condition to drive, with good brakes, steering and tyres.

OPPOSITE This 3C Mk III was built in 1978 and restored by the author in 2013. INSET The same machine before restoration. It was purchased new by Lewisham Borough Council in 1978, originally priced at £19,500 ex-works. In 1990 it was auctioned off in Merthyr, South Wales, and spent 13 years doing light work on a farm plus snow clearing in the winter.

RIGHT The machine is seen here in 2013, fitted with the original hand-held breaker on the side and the Jaw Bucket on the excavator.

It goes without saying that any 3C Mk III will be almost 40 years old, so the machine won't operate as if it's new and will be less powerful and less comfortable to use than modern-day machines. However, if it has been serviced and well maintained throughout its working life there will still be plenty of life left in it. But still, the occasional breakdown should be expected, with hydraulic hoses being one of the most common issues with a machine of this age.

Typically old machines end up on farms, where they're used periodically for clearing drainage ditches and remedial work on the land and are left standing outdoors in all the elements. The cabs are prone to corrosion and rotting away around the base where they sit on the steel fenders, and this can lead to the window rubbers, which hold the glass in place, coming away and the glass becoming loose. This then allows the rainwater into the cab, causing corrosion to the inside and to the seat. If the corrosion becomes so bad that the cab frame loses its structural integrity, the cabs are often removed and scrapped, leaving the cab open to the elements and making it particularly unpleasant to use in inclement weather.

Choosing the right specification

If restoration is your heart's desire, then you should consider if there are any particular features you'd like on the machine, such as a 6-in-1 shovel on the front and the more unusual extending dipper or jaw circuit on the rear. The best 3C Mk III models were the later Powertrain models with hydraulic brakes, which provide better braking performance, and with torque convertors that function better than the earlier Brockhouse versions. However, Powertrain models are harder to find as they were only in production for a short period. If you're looking for a high-specification machine with features such as the tool circuit and jaw circuit, these are rare, as at that time the concept was in its infancy and they were expensive options.

Where to look for a machine

There are several places to look for a JCB: on-line auction sites, plant and farm auctions and advertisements in machinery magazines, which can be found in most newsagents. For

RIGHT The pipework running down the dipper operates the Jaw Bucket (not fitted) and is very rare to find on a machine, as it was a very expensive option in its day.

auctions, first check when the pre-sale viewing is taking place, as this is designed to allow an opportunity to examine the machine and to check whether it runs or not before the sale. If the machine is incomplete, or in poor condition, make a note of which parts are missing and check how easy they'll be to find, and the costs.

If the machine is part of an online auction the pictures can sometimes be misleading, and most sellers will provide a contact telephone number to make further enquiries. It's strongly recommended that you take the opportunity to visit the machine to assess its condition and test its operation. It's important to check with the seller whether they're selling through a company, as it may qualify for tax on top of the sale price. If you intend to take the machine on the road, it would save a lot of time if the machine has a logbook to enable you to register it for road use.

What to look for when buying

Before starting the machine, take a good look around it and check the condition of the tyres, that they hold air and have sufficient tread. Look under the rear of the machine around the rear axle and look for signs of oil leaks where the axle fits on to the wheels, as old machines are prone to leaking from their rear wheel hubs.

Checking the engine

Check the engine oil and transmission oil dipsticks to ensure there's sufficient oil. Also look at the condition of the oil. If the engine oil is very thick and black, it's overdue a service. Neither the engine nor the transmission oil should be grey or emulsified as this would indicate that water is present and suggest that a rebuild is inevitable to avoid total failure of the unit.

When starting the machine smoke may be present at first, which isn't unusual, but it should clear. If the engine continues to smoke there may be a problem with worn pistons

ABOVE The standard shovel on this 1977 3C Mk III is in very good condition (*ie* not bent!) and has a good set of teeth for digging in hard ground.
(Neil Green)

LEFT A clean engine bay makes it easier to spot leaks from the engine. Look for any signs of water leaks from the cooling system and any water or oil leaks from the head gasket.

and rings or with the fuel injection system, so this may require further investigation. Look for any signs of oil or water leaking from around the cylinder head area, which would indicate that the head gasket is blowing and in need of replacement. The water in the radiator shouldn't have any signs of oil and should sustain a constant level.

Hour clock

Although the machines are fitted with an hour clock on the dashboard, it shouldn't be assumed that it's working, as the driveshafts on the cable often fail, which means the clock counter stops working. On a machine of this age it wouldn't be unusual for it to have worked in excess of 10,000 hours or much more, but if a machine shows genuine hours of less than 6,000 then this would be considered low for its age.

Checking the steering

With the engine running, the steering should feel light and responsive in both directions, without any hydraulic pump whine when turning. If the steering feels jerky or heavy, this would suggest that the hydraulic filter for the steering is blocked or the orbital valve or pump is worn. The steering pump and orbital valve can be serviced, but replacements are very expensive. It's always good practice to raise the wheels off the ground using the front shovel, then hold the front wheel and rock from side to side to feel for excessive play in the wheel hub bearing and/or stub axle, which may be visible on the pin and bushes if excessive.

Checking the transmission

Next is to evaluate the condition of the transmission. If it has a manual gearbox, check that the machine drives in all the gears and in both high and low ratio, as faulty gearboxes are very expensive to repair and can cost more than the machine. The clutch should be smooth and have sufficient pedal travel before 'biting'. If the machine is fitted with a torque convertor, check the functionality in both directions and that the machine takes up drive. The earlier Brockhouse versions have a slight delay after moving the directional lever into position before taking up drive in both forward and reverse, so this is normal. On torque convertor models use the dipstick to check the level and condition of the oil.

Checking the brakes

Check the braking effort and stopping ability of the brakes, which are usually poor due to lack of service and adjustment. Both the foot and park brakes should hold the machine on a hill, but will probably need to be serviced. It may be a case of applying lubrication oil to all the linkages and adjusting the brakes to take up the slack due to wear on the discs and/or linings.

If the machine has the earlier calliper-type brakes, these can be replaced fairly easily by first removing the cab floor plates to access the brakes system. Replacement parts aren't too expensive.

Checking the hydraulics

With the machine running, operate the front loader and excavator and cycle the rams several times. Then check each ram for signs of leaks and worn wiper seals. If the piston rods have signs of pitting or rust and are leaking sufficiently that you consider new seals are needed, then the piston rod should also be replaced to prevent damaging the new seals.

The loader and excavator levers should be checked for wear and leaks from the tops of the spools where the levers connect. The O-rings and wiper seals on the top of the valve block can be replaced without removing the block,

BELOW The main gauges are positioned on the right-hand side of the front console, but don't rely on the gauges still working as the engine could overheat and suffer from low oil pressure, which is the sign of a worn engine.

as can the levers and ball joints on the top of the spools if there's excessive wear. Replacing the levers can make a significant difference to the operation of the machine, as it significantly reduces the amount of effort and hand movement required, as well as improving the overall control.

A very simple test to give you an indication of the hydraulic pump's condition is to raise the front loader with the engine on tick-over. The loader arms should raise at a constant speed without the engine stalling or the pump whining excessively. If the loader arms move at a very slow speed or not at all, it means that the pump is worn. Resealing or replacing the pump can be expensive.

To evaluate the machine's hydraulic power on the excavator, position the bucket flat on the ground and check that the machine has sufficient power to lift itself off the ground at around 1,000rpm. If the machine's incapable of lifting its own weight there are several likely causes, including a worn hydraulic pump, leakage of the spool in the valve block and worn ram seals, but a most likely cause would be a worn auxiliary relief valve (ARV), which may just need to be adjusted.

Hydraulic hoses

Hydraulic hoses on old machines are prone to failure due to their age and fatigue resulting from years of flexing. Failure is often due to the hose becoming snagged or crushed as a result of being incorrectly routed from a previous repair or replaced with an incorrect-length hose, which allows them to rub against another hose or the machine, causing them to wear away their outer protective cover. The exposed wire braid then corrodes and breaks, rupturing the rubber hose it covered.

Checking pivot pins and bushes

Check the condition of the pivot pins and bushes for signs of excessive wear, particularly on the excavator. This can be done by pushing against the boom, dipper and bucket linkages to assess the amount of movement. Replacement bushes are still available and are relatively inexpensive, but it can require excessive effort to knock the pins out, and in extreme cases a hydraulic press might even

ABOVE The excavator valve block and hoses are concealed by a fibreglass cover that's easily removed to carry out repairs but is often not refitted and consequently missing on old machines.

LEFT Fitting incorrect-length hoses and poor routing around the excavator kingpost can lead to wear and premature failure of hydraulic hoses.

LEFT Look carefully for any cracks around the pivot points and at the points where the castings join the fabrication. The degree of movement when pushing hard against the excavator arm will indicate the amount of wear in the pins and bushes.

RIGHT The gauze filter at the bottom of the fuel tank is often forgotten and never cleaned, so is worth checking to avoid fuel blockages.

be needed if they're tight in the bore. Look carefully around the ends of the pins for signs of welding, as loose pins may have been welded in place to prevent them from rotating, which creates difficulty in removing them.

Structural failures

Look carefully for any signs of structural failure to the chassis, especially around the loader tower area and on each side of the chassis members running along the side of the engine. As the side members contain hydraulic oil, any cracks in the Hydrachassis would be the source of leaks and can be difficult to weld. Other areas to check for fine cracks or remedial welds and reinforcement plates are around the top of the boom and dipper. If repaired properly they shouldn't be a cause for concern, but should be inspected periodically to ensure the repair is lasting.

Cleaning the fuel tank

Finally, if you intend to put the machine to work it's strongly recommended that once you've bought it the machine is fully serviced first by replacing the engine oil and filter, air filter and fuel filters. If the diesel sediment bowl has excessive dirt and/or water, then it's recommended that the fuel is drained and washed through with clean diesel. If the tank is particularly dirty a pressure washer can be used to thoroughly clean it. It can be difficult to fit a jet wash lance though the fuel filler spout due to the elbow on the filler neck, so the most effective method is to pass it through the large round hole at the top of the tank in the centre, where the fuel sender float is fitted. There's also a gauze filter that fits into the union which screws into the underside of the tank, to which the fuel feed pipe connects and runs to the lift pump. Access to this union is particularly difficult as it's in the end of the tank where all the hoses are routed to the loader valve block, so in most cases it's forgotten about or deemed too difficult to access. If there's a lack of fuel coming from the tank a blocked gauze will be the source of the problem.

Restoring a machine

It's important to understand that the restoration of a JCB isn't for the faint-hearted, not just because of the level of work involved but because it can also be very costly, running into tens of thousands of pounds, to get a machine back to 'showroom condition'. Despite this, it's

RIGHT The restored 3C Mk III was put through its paces after restoration at the JCB demonstration ground located in Ashbourne, Derbyshire, which was a World War Two airfield.

ABOVE The restored 3C Mk III owned by the author parked next to a modern 3CX. The yellow colour on the 3C was a lot more lemon than on the modern machine.

immensely rewarding to see a machine brought back to its former glory and then taking it to vintage shows for others to enjoy.

Machine storage

The first obvious question you'll need to ask yourself is where you will store the machine during restoration and how accessible it will be when you intend to carry out the work, whether this be in the daytime, evenings or at weekends. A local farmer may provide undercover storage in return for a small fee or a local yard may provide space in a corner. Consideration needs to be given to carrying out the repairs in a workshop since lifting equipment will be required at times to remove heavy components.

Necessary skills

If you don't have a mechanical background or previous experience of working on other types of commercial vehicles the restoration is obviously going to take a lot longer to do, and it'll be necessary for you to purchase tools along the way. A broad range of skills are required to carry out a full rebuild, ranging from mechanical and hydraulics through to fabrication, electrical and bodywork repairs. Even the most experienced of restorers don't necessarily have the skills and equipment to do everything themselves and will call upon assistance for more specialist tasks, such as spray painting.

Setting a budget

You'll need to set yourself a budget for the project, as costs can quickly escalate, and from my experience repairs typically cost more than first expected. If you know someone who's restored another vehicle it would be helpful to discuss your project with them and get some idea of how much the work costs to carry out. By far the cheapest option will be to do as much of the work as you can by yourself to keep the labour costs to a minimum. Also, you could join a local vintage club, which should be able to put you in touch with people possessing some of the skills that you need.

Sourcing parts

Once you have a machine you will be eager to get the tools out and start stripping it down. Before you do anything, you should spend time planning the project and locating the parts that you have identified you will need. Shortlist companies that have the necessary experience and skills to carry out the work that you're unable to complete yourself. The Internet is always a good place to source parts, in particular online auction sites. If you come across rare parts that you require, purchase them whilst you can, as you may not find another within the timescale in which you wish to complete the restoration. In all of the restorations in which I have been involved I have located parts from other machines or even purchased a second machine to act as a donor for parts to aid with the restoration. Don't underestimate how long it can take to locate hard-to-find parts or commission parts that need to be specially made. In some cases

LEFT A full set of new parts ready to go into the refurbished cab. New cab parts are no longer available from JCB, which makes finding replacements considerably more difficult.

drawings will have to be produced to enable the parts to be manufactured.

Stripping down the machine

Establish what you're going to do with the parts once they have been removed from the machine – are they going to be sent off for repair or refurbishment or kept in temporary storage? If you are doing the work yourself, do you have the necessary tools, parts and knowledge to carry it out? It's always easier to approach the work in manageable bite sizes in order not to lose parts or forget how they fit back together, so take plenty of pictures! Try and keep the machine running and mobile for as long as possible. This ensures that the machine can be moved into different positions to facilitate the removal of other parts.

If you are setting out to achieve a concourse restoration, it'll be necessary to remove the loader and excavator and to strip the machine right down to its bare chassis to enable every part to be properly cleaned, shot-blasted, filled and painted. The skid unit, *ie* engine, transmission and axle can be kept complete and painted.

Sheet metal work

The steel fenders often become damaged, bent and, in many cases, rotten right through, particularly where the cab sits on its fenders, as the foam on which the cab sits holds moisture. If the steel is in a poor condition it can be cut out and replacements fabricated, which will require a large hydraulic press to form the 90° angle.

CENTRE Having been shot-blasted, the main parts are taken back to the workshop to be undercoated and top-coated. It's essential to get the bare metal primed immediately after shot blasting to prevent it from oxidising and rusting.

LEFT Due to the corrosion being so bad it was necessary to replace the entire wings, using the old ones as a template to fabricate the new ones.

Because of the work they do JCB cabs are notorious for corrosion and rust, particularly around the base and framework, which sits on top of the fenders. To fully appraise the extent of corrosion it helps to shot-blast the cab frame first, as this makes it easier to see what can be filled and which sections need to be completely cut out and replaced. Replacing sections of the cab pillars is particularly complicated due to the compound curvatures. A lot of skill is required to make a seamless joint. This may be one job for which you will need to employ the expertise of a skilled craftsman who has the necessary tools.

Cab glazing

The cab glass is held in place using 'Claytonrite'-type rubber seals. A special tool is needed to fit the filler strip. Although this type of rubber seal is no longer used on modern equipment it's still available from most machinery glaziers.

Electrical equipment

The genuine electrical components, lights, dashboard gauges and dials are now obsolete

ABOVE The cab frames suffer corrosion, particularly around the base. A number of sections on this cab have been cut out and replacement sections formed and welded into the original frame.

BELOW The front console is made from rigid urethane material, which once dismantled and cleaned can be sprayed black to make it look like new again.

and are particularly difficult to find. It may therefore be necessary to settle for good-condition second-hand parts. The table below lists all of the original equipment to try and source:

Component	Manufacturer	Type
Front side lights	PMG (Britax)	1518
Rear side lights	PMG (Britax)	1518
Headlights	Lumax	
Work lights	Lumax	
Beacon	PMG (Britax)	347:
Interior light with switch	PMG (Britax)	791:00
Wiper motors	SWF	
Wiper arm, front	PSV Wipers	
Wiper arm, rear	PSV Wipers	
Wiper blade	PSV Wipers	
Warning lights	Arco electrics	
Ignition switch	Lucas	
Indicator stalk	Lucas	
Flasher unit	Lucas	8FL 36.35048B
Interior mirror	Wincard	
Door catch	Jagger Ltd	
Door handle	Jagger Ltd	01/5459
Mirrors – 10in x 8in	Raydot	3202
Mirror arms	Wincard	980.585
Heater panel switch	Lucas	39823 159SA
Panel rocker switches	Lucas	30829A
Oil temperature gauge	AC Delco	7966216
Number plate light	PMG	840:01

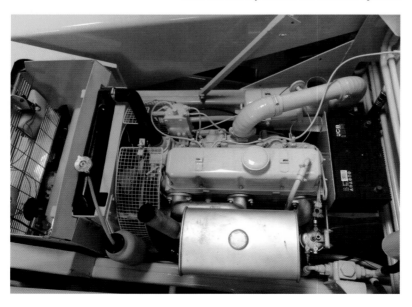

BELOW When spraying the engine bay, remove the radiator and as many parts as possible to get good paint coverage around all areas. Use special heat-resistant paint on the exhaust system.

Cleaning the parts

Before parts are shot-blasted they need to be thoroughly cleaned. The best way to achieve this is by using a hot water pressure washer and applying a degreasing agent such as JCB Plant Wash, which will help to remove years of grease and oil. Any burred edges should be removed using a grinding disc and pitted metalwork needs to be filled and sanded back to leave a smooth finish in preparation for painting. From my experience it's better to fit the metal JCB nameplates once the main machine is painted, to ensure that the surfaces behind the nameplates are painted. This prevents future corrosion, and although the original plates were welded to the boom and fenders it is recommended that a strong adhesive is used to bond them in place, as welding them on damages the paintwork and would be a source of corrosion in the future.

Paint

Although the shade of yellow has changed slightly through the decades, it's been the primary company colour ever since the introduction of the Hydra Digga. The red wheels and shovel were dropped in the early 1980s. In 1972 the cab colour was changed from yellow to a pure white to help reduce heat radiation inside. This continued through to the 3CX model until the early 1980s, when a black cab was introduced. Backhoe loaders have always been spray-painted by hand and were originally fully assembled, finished in primer then washed and masked up prior to application of the top coat.

The paint used at the time contained lead, which is no longer allowed. The original yellow used was 'lemon chrome', but a different shade was used on the 3C Mk IIIs. This was unique to JCB but didn't have a name. 'Ayres red' was used on the shovel, wheel rims, attachments and the buckets for a period before changing to yellow. The white cab is BS5252 00E55, 90–94% gloss level.

Care and safety

1 **CHECK** lifting equipment before use.
2 **ALWAYS** face the machine when descending from the cab.
3 **DO NOT** carry passengers.

4 **DO NOT** walk under the arms.
5 **DO NOT** drive near the edge of a hole or trench.
6 **DO NOT** lean out of the rear window.
7 Keep loader shovel about 457mm (18in) above the ground when:
Travelling with a full shovel.
Reversing up a gradient.
Travelling across a gradient.
8 **DO NOT** operate controls from outside the cab.
9 **SECURE** all moving parts of the machine when not in use.
10 **AVOID** parking on slopes.
11 **ENSURE** brake pedals are latched together before travelling on public roads.

Routine maintenance

Grease points around the machine had to be greased at the following intervals:

Loader
- Grease points – every 10 hours.

Excavator
- Pivot points – every 10 hours.
- Check actuator level using the dipstick every 50 hours. ('A' = filler and dipstick; 'B' = bleed-off pipe.) Leakage may indicate a seal failure.

Extending dipper
- Grease sliding faces every 50 hours.

Pump drive
- Grease universal joints every 50 hours.

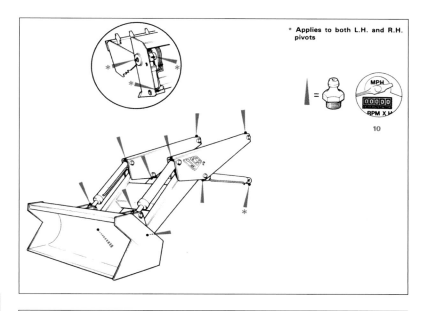

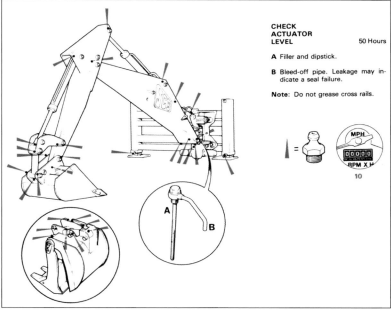

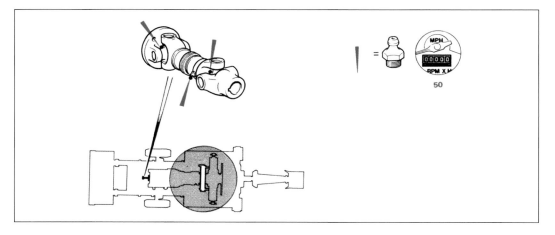

TOP Loader end greasing points.

ABOVE Excavator end greasing points.

LEFT Pump drive greasing.

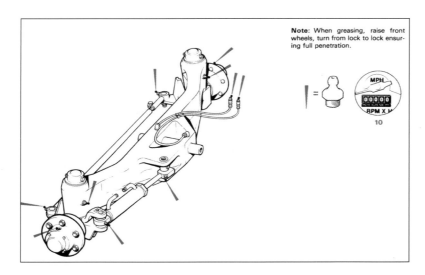

Note: When greasing, raise front wheels, turn from lock to lock ensuring full penetration.

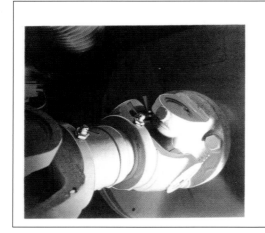

TOP Front axle greasing.

ABOVE Propshaft greasing (Powertrain versions only).

RIGHT Engine oil filter.

FAR RIGHT Fuel system.

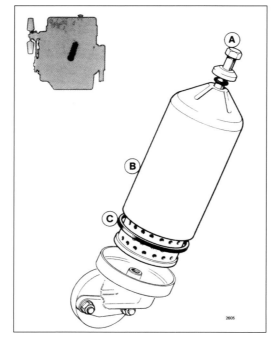

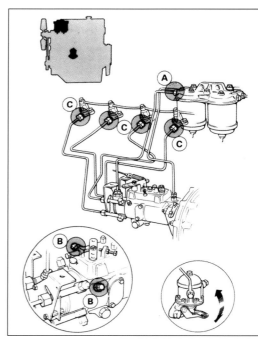

Front axle
- Grease all nipples every 10 hours. (Note: When greasing the axle, raise the front wheels and turn from lock to lock to ensure full penetration.)

Propshaft
- Grease every 50 hours.

Engine oil
- Check level every 10 hours.
- Drain and refill every 200 hours.

Engine oil filter
- Renew element every 200 hours.
- Unscrew bolt (A) and remove body and element (B). Ensure all metal parts are clean and reassemble and seal (C). Smear new seal with oil before assembly.

Bleeding the engine fuel system
If the engine fails to start or misfires, or if any part of the fuel system has been disconnected, the system should be bled as follows: Slacken screw (A) and operate lift pump priming lever until air-free fuel flows from screw, then retighten. (Note: If the lift pump priming lever can't be operated, turn the engine until it can be moved.) Then slacken screws (B) and operate the lift pump lever

until air-free fuel flows from screws, then retighten. Next, slacken unions (C), open throttle, push in engine stop control, operate starter until air-free fuel flows from unions then retighten.

Engine fuel filter

- Drain every 50 hours and renew every 400 hours.
- Drain off any water in the glass sediment bowl by turning tap (A).
- Renew elements by supporting bowl (B) while unscrewing bolt (C). Remove bowl and element (D) using a twisting motion. Reassemble using new element and sealing rings. Finally, bleed the fuel system (see fuel bleeding instructions above).

Engine fuel sediment bowl

- Drain every 50 hours.
- Drain off any water in glass bowl by turning tap (A).
- Clean – periodically dismantle by clamping the inlet fuel line to prevent loss of fuel. Support the bowl (B) and unscrew bolt (C) and wash the element (D) in clean fuel. Reassemble in reverse order ensuring the gaskets are in position and the bolt isn't over-tightened. Finally, remove the fuel line clamp and bleed the fuel system.

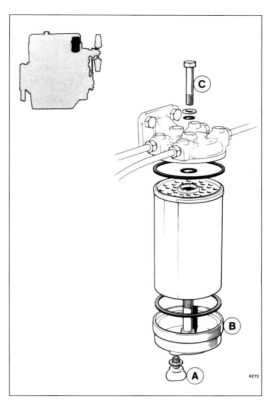

LEFT Fuel filter.

Engine fuel lift pump

- Clean gauze every 400 hours.
- Unscrew bolt (A) and remove cover and gauze (B). Wash gauze with petrol and a soft brush and reassemble, tightening bolt just sufficiently to make a seal. Finally, bleed the fuel system.

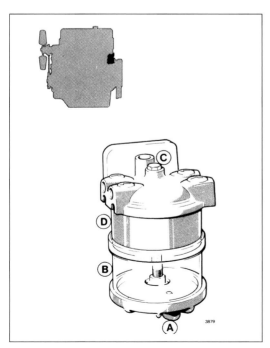

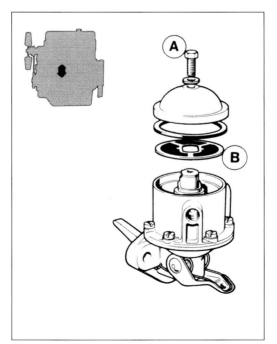

FAR LEFT Fuel sediment bowl.

LEFT Fuel lift pump.

95
THE OWNER'S VIEW

RIGHT
Engine cooling system.

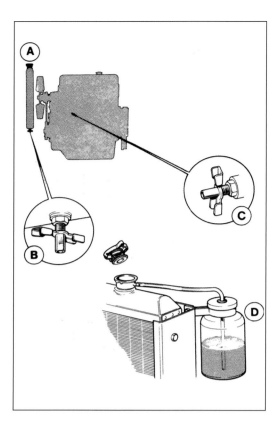

BELOW
Engine fan belt.

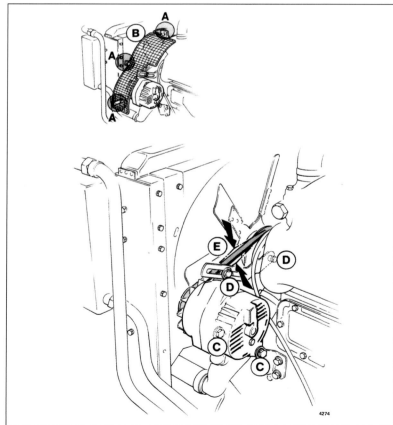

Engine fuel injection pump
- Check oil level every 50 hours.
- Drain and refill every 200 hours.

Engine cooling system
- Check level every 10 hours. A = radiator filler cap; B = radiator drain tap; C = cylinder block drain tap; D = expansion bottle.

WARNING: Do not remove the radiator cap when the system is overheated. Keep the expansion bottle half full. Top up the system with an appropriate solution of anti-freeze and water.

Anti-freeze
- Add 0.15 litres (¼ pint) to coolant in expansion tank. Add to radiator (see table below).

	Anti-freeze	Starts to freeze
25%	3.5 litres (7.2 US pints)	*-12°C (10°F)
57%	8 litres (16.2 US pints)	*-30°C (-22°F)

Engine fan belt
- Adjust every 50 hours. For easier access to the alternator, unscrew bolts at A to remove the fan guard (B). Slacken bolts C and D. Move the alternator away from the engine until there's 13mm (0.5in) slack at E. Retighten bolts C and D, tightening bolt C at the rear of the alternator last. Finally refit the fan guard and secure.

CAUTION: Any leverage required to position the alternator must be applied at the drive end bracket *only*, using a wooden lever.

Engine air filter
- Clean the vacuator valve every 50 hours.
- Renew element every 800 hours.
- *Do not* disturb filter elements between service intervals unless the filter warning light shows red.
- *Do not* tap or knock the element as this can dislodge debris contained within the filter.
- Ensure that the seal at C is smeared with grease and is seated correctly.
- Check the tightness of induction hose clips at regular intervals.

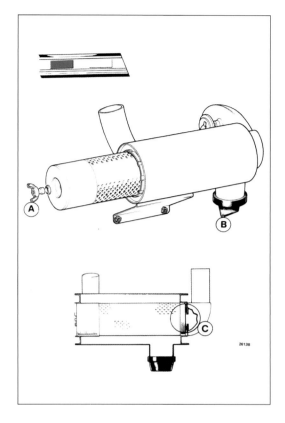

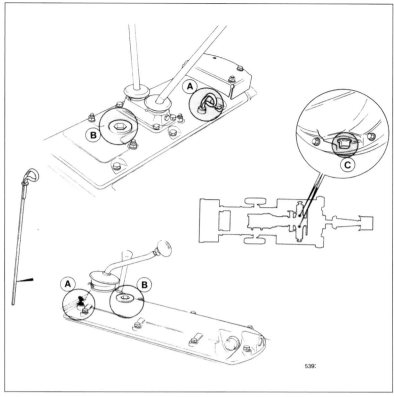

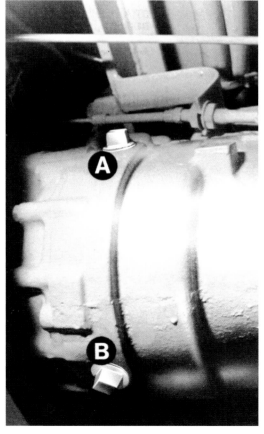

Removal of air filter

Unscrew the wing nut (A) and withdraw the element. Blow dirt from the element using compressed air and check for perforation by placing a strong light inside the element. Renew if perforated. Remove and clean the vacuator valve (B).

Gearbox oil

- Check level every 50 hours.
- Drain and fill every 800 hours.
- When checking the oil level, park the machine on level ground. A = dipstick; B = filler plug; C = drain plug (one each side).

Powertrain models

- To check the level, remove fill/level plug (A).
- To drain and refill, park the machine on level ground. Remove fill/level plug (A) and drain plug (B). Refit drain plug (B) and refill at A.

Torque convertor oil level

- Check level every 10 hours.
- Run the engine for two minutes then, with the engine idling, check the oil level (dipstick A).

ABOVE LEFT Air filter.

ABOVE Gearbox oil.

LEFT Gearbox oil level – Build /5.

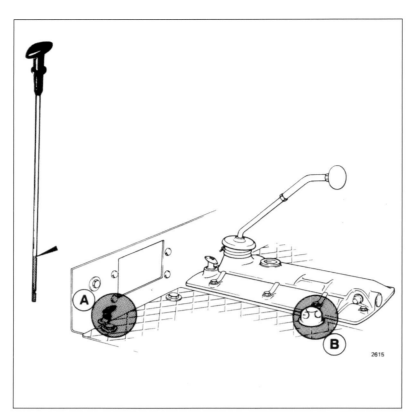
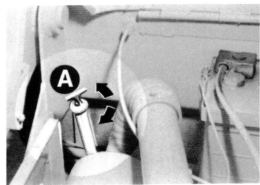
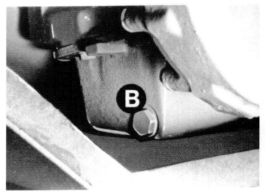
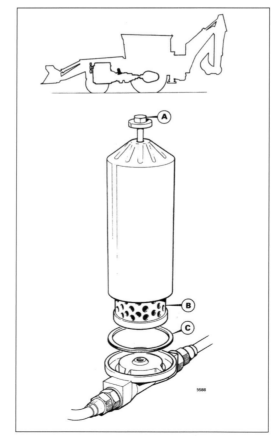
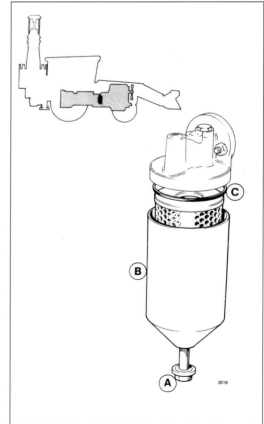

ABOVE Torque convertor oil level.

ABOVE RIGHT Torque convertor oil level (Powertrain).

RIGHT Torque convertor oil filter (Powertrain).

FAR RIGHT Torque convertor oil filter – Build /3.

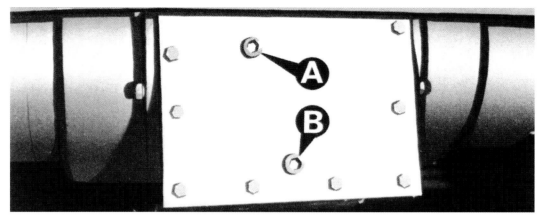

LEFT Rear axle oil level – Build /5.

- Drain and refill every 800 hours.
- Drain (drain plug C) whilst the transmission is hot.
- Top up or fill (filler cap B) with transmission fluid (see 'Lubricants' listing on page 113).

CAUTION: Do not overfill.

Powertrain models
- Check level daily.
- Drain and refill every 800 hours. A = dipstick/filler; B = drain plug. Note: Twist the dipstick to remove.
- Check the oil level immediately after stopping the engine.

Torque convertor oil
- Renew the element every 400 hours.
Unscrew bolt (A), remove body and the element (B). Clean all metal parts, then reassemble with new element and seal (C).

Powertrain models
Unscrew bolt (A), remove body and element (B). Clean all metal parts and reassemble with new element and seal (C).

Rear axle oil level
- Check level every 50 hours.
- When checking, park the machine on level ground.
- Remove fill/level plug (A). Oil should run from hole. Top up if necessary.
- Drain and refill every 800 hours.
- To drain and refill, park the machine on level ground. Remove fill level plug (A) and drain plug (magnetic) (B).
- Refit drain plug (B) and refill at A (see lubrication chart). Wipe drain plug clean before refitting.
- When filling the axle, allow time for the oil to drain past the axle shaft bearings into the axle arm before rechecking the level, otherwise a low level will result.

Clutch adjustment
- Check clutch greaser every 10 hours (half turn).
- Keep clutch greaser filled with grease (see lubrication chart).

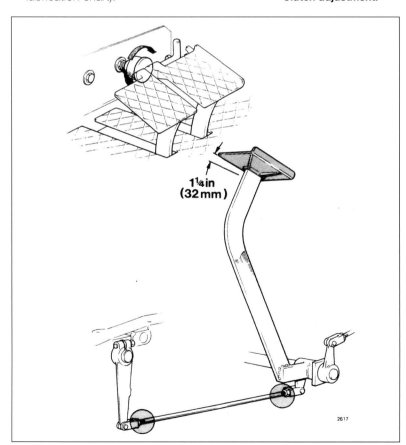

BELOW Clutch adjustment.

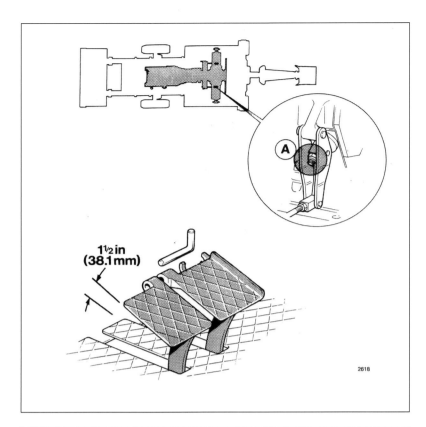

LEFT Brake adjustment up to machine 129502.

- Check free movement every 50 hours. Adjust length of rod to obtain 32mm (1¼in) of free movement at the clutch pedal.

Footbrake

Up to machine serial number 129502

- Check adjustment every 50 hours.
- Left foot brake – screw up nut A to obtain 40mm (1½in) free movement on left-hand pedal.
- Right foot brake – adjust until right-hand pedal lines up with left-hand pedal when free movement is taken up.
- The machine should brake in a straight line when the brake pedals are latched together. If the machine pulls to one side slightly, increase free movement on the pedal operating that side.

From machine serial number 129503 (skid builds only)

- Check adjustment every 50 hours.
- Adjust the brake linkage to the dimensions given between the clevis pin centres. Set the handbrake stop (A) to limit forward travel of the handbrake lever at the second notch of the rack. Jack the rear wheel off the ground and tighten each adjusting nut (B) until the wheel just locks, and then back nut off ten flats. Road-test to check the brake balance, backing off the appropriate nut, if required, to adjust balance.

Powertrain models

- Remove cover A, unscrew B and top up if necessary to correct level mark. See the lubrication chart for correct fluid. *Do not allow fluid to fall below 'DANGER' mark.*
- Check adjustment every 50 hours (dry brakes only).
- Jack the rear wheels off the ground. Tighten each adjusting nut (C) until the wheel just locks and then slacken off 1.5 turns. Road-

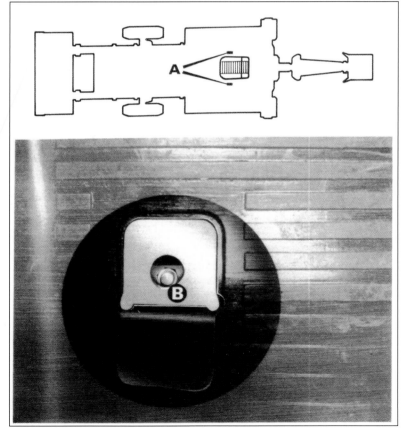

LEFT Brake adjustment.

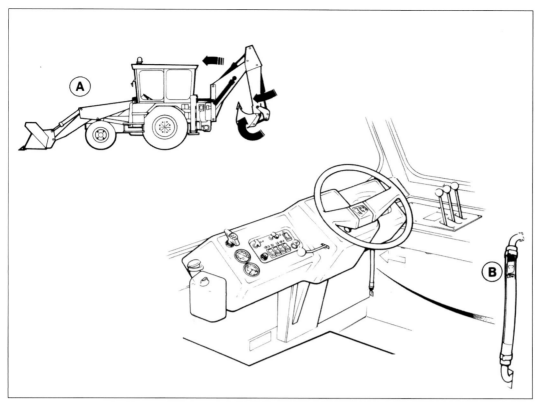

LEFT Hydraulic fluid level.

BELOW Parking brake (Powertrain).

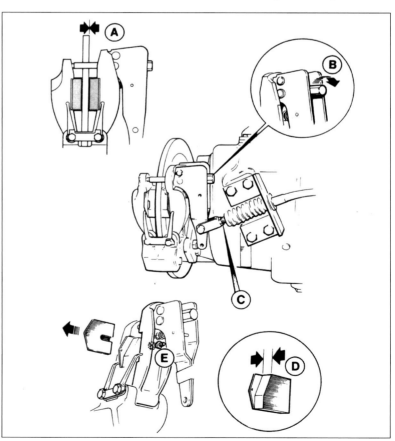

test to check the brake balance. To adjust balance, back off the appropriate nut and retest the machine.

WARNING: Use of incorrect brake fluid will cause serious damage.

Parking brake
(Powertrain models)
- The pads should just touch the discs as at A. Adjust at nut B if required. Take up any slack at C.
- Renew pads when thickness D of friction material is 3mm (0.125in).
- Back-off all adjustment, slacken nut E and withdraw pads. Install new pads and adjust.

CAUTION: The parking brake *must not* be used to slow the machine from travelling speed, otherwise pad wear will be excessive.

Hydraulic fluid
- Check the level every 10 hours.
- Drain and refill every 1,600 hours.
- Fluid level – position machine as shown at A with the front shovel resting on the

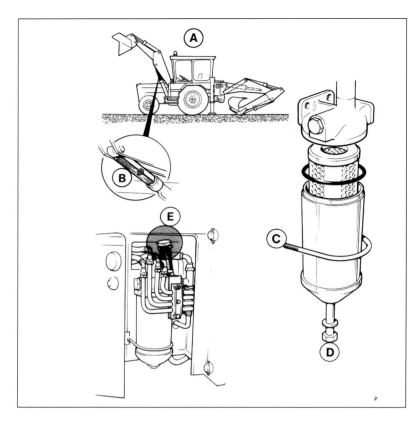

ground, boom ram closed and dipper and bucket rams open. The correct level is then two-thirds up the level indicator tube (early machines) or between the two indicator marks as shown at B (later machines).

Hydraulic filter

- Renew the element every 400 hours.
- Position machine as at A with all rams extended to prevent siphoning when the filter body is removed. Install ram prop B before attempting to change the filter element.
- Remove U-bolt C, unscrew through bolt D until the filter body can be withdrawn approximately 25mm (1in).
- More than 5 litres (1 UK gallon) will flow from the filter at this stage. Remove bolt D and filter body, discard element and seals. Wash all metal parts in clean JCB 'Special' hydraulic fluid. Reassemble using new element and seals. Top-up the hydraulic system at E; alternatively the oil can be poured directly into the hydraulic tank by removing the air breather cap on the top of

ABOVE Hydraulic oil filter.

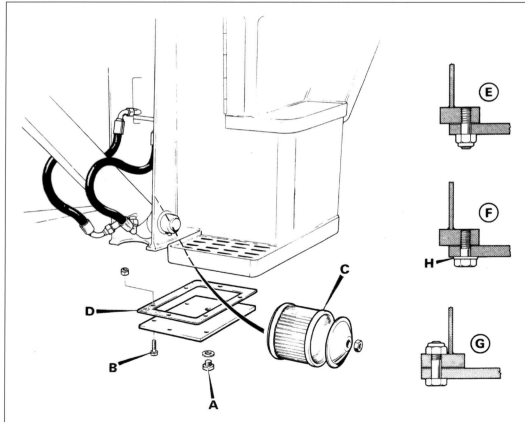

RIGHT Main suction strainer filter.

RIGHT Steering pump strainer.

the tank in front of the windscreen.
- Note that the filter body is heavy, and assistance from a second person is recommended.

WARNING: Do not attempt to refill the hydraulic system through E whilst the engine is running, as hydraulic oil will exit at high pressure through the filler cap.

Main suction strainer
- Clean every 1,600 hours.
- Remove plug (A) to drain the hydraulic tank and then unscrew bolts (B) and remove access plate. Remove the strainer (C) and wash in petrol or parts-washing chemical. Reassemble using new gasket (D).

NOTE: Three alternative methods of access plate attachment are shown at E, F and G. If F is used, renew the sealing washers.

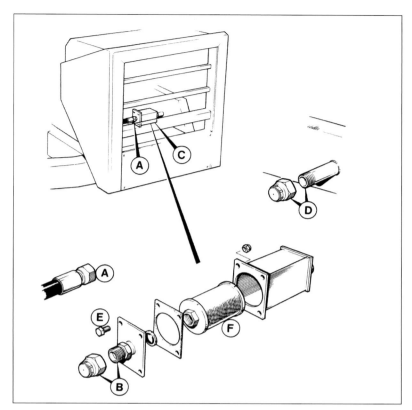

Steering pump strainer
- Clean the element every 1,600 hours.
- Disconnect and blank hose (A) and connection (B). If the hydraulic oil isn't drained, unscrew the complete strainer assembly (C) and blank stud pipe (D) to avoid excessive fluid loss. Unscrew bolts (E) and remove element (F) and wash in petrol or suitable parts cleaner before reassembling using a new seal and gasket.

CAUTION: Tighten bolts (E) to a torque setting of 15–20lbf/ft (2.1–2.8kgf/m). Over-tightening will cause distortion of the end plate and subsequent fluid leakage.

Headlamps
- With the lamps on main beam, the light pattern should fall as shown in the picture.
- To adjust, slacken A or B and reposition the lamp body as required.

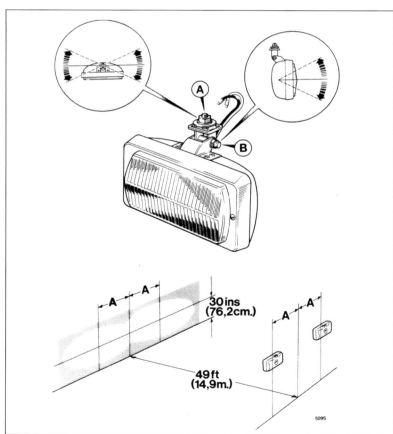

RIGHT Headlamp adjustment.

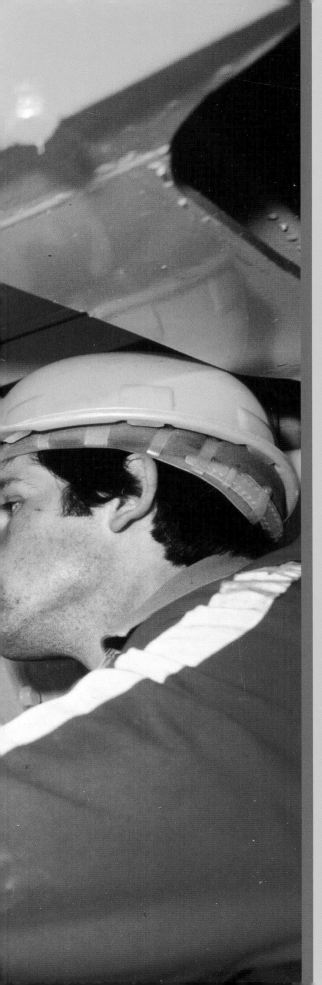

Chapter Seven

The mechanic's view

The 3C MK III is straightforward to service, and in order to keep a machine working efficiently it is essential that it is properly and regularly maintained. The Leyland skid unit is basic and is sufficiently old-fashioned to allow it to be maintained by the average home mechanic. Due to the age of the machine, the hydraulic components are normally the main source of problems due to leaks or damage, in particular corroded cylinder rods and worn hydraulic seals.

OPPOSITE The park brake is mounted on the propshaft on the Powertrain versions.

Safety first

Secure all moving parts of the machine to prevent inadvertent actions by other personnel causing dangerous movement and possibly injury. For example, always set bucket and shovels on the ground or on adequate supports before commencing work.

Never work beneath a machine unless it's standing on hard ground, preferably concrete or tarmacadam. If a machine is lifted to improve accessibility, always place substantial supports under the mainframe before work is commenced.

Never take risks. Always check lifting equipment, slings and chains etc before components are hoisted. Always use equipment that's adequate for the job in hand.

Hoses and pipes

Always fit blanking plugs whenever hoses or pipes are disconnected. Serious damage can occur if dirt or abrasive material enters the hydraulic system.

Rest the digging equipment on the ground, stop the engine and operate all control levers to vent any residual pressure from the hydraulic circuit before any hoses or pipes are disconnected. This minimises the chance of injury from pressurised spurting oil when a connection is loosened.

Unsupported hoses or pipes must not touch each other or adjacent components or chafing will be caused.

Hot oil burns!

Carefully test the temperature of adjacent pipework and components with your hand prior to disconnecting a hose or pipe. If the oil is considered too hot, allow sufficient time for it to cool before the connection is loosened. Renew pipes if any of the following faults are found:

- Core damaged or distorted.
- Thread damaged.
- End-fitting damaged or leaking.
- Outer surfaces damaged or severely corroded.

When renewing hoses or pipes, loosely connect each end, settle the hose or pipe into position and tighten the connections. Clamps should be tightened sufficiently to retain the hose or pipe without crushing it. If a hose is connected to moving components, the components should be moved over the full extent of their travel to check that the hose doesn't foul them as it's bent.

CAUTION: Hoses must *never* be twisted when fitting.

Dismantling and assembly

Thoroughly clean the exterior of all components before dismantling. If dismantling is commenced without removing a component from the machine, cleanliness is doubly important since the service life of a component is drastically reduced if dirt or abrasive matter isn't excluded at all times.

Clean it first!

Sometimes components are dismantled in adverse conditions without workshop facilities being available. If this is necessary suitable precautions must be taken to prevent dust and dirt particles getting on internal parts, particularly on oiled surfaces where it'll be retained and contaminate any new oil with which a component is filled.

Clean it carefully!

Select your cleaning agents and solvents with care.

WARNING: Solvent-type cleaners are dangerous when used in a confined area. Avoid inhalation of the vapour or contact with the skin. Don't use near a naked flame.

Take care!

Renew O-rings, seals and gaskets – regardless of condition – whenever they're disturbed, unless instructed otherwise. Lightly coat O-rings and gaskets with hydraulic oil before fitting, unless instructed otherwise. Good sealing pays, so coat gaskets with a suitable jointing solution before fitting, unless instructed otherwise.

Bearings

Don't remove bearings unless they are to be renewed or it is unavoidable in a dismantling sequence. Whenever possible use a suitable puller or extractor to minimise the risk of damage to both the bearing and any adjacent parts. Keep bearings with loose rings, bails or rollers as matched assemblies; they are selectively assembled and parts don't interchange from one bearing to another.

Inspection

- Check carefully that the bearing has had ample lubrication and that the lubricant hasn't been contaminated.
- Check for discolouration and signs of overheating.
- Check the bearing for excessive clearance and the cage for wear, cracks and scuffing. If in doubt regarding the clearance, compare it with an identical new bearing. (Note that journal bearings are graded by internal clearance when new. This is indicated by one, two or three faint circles marked on one ring. Ensure that both bearings are of the same grade when comparing new with old.)
- Check the finish of the locating surfaces (A and B). They may have been lightly marked on assembly but shouldn't be polished. Polishing indicates that the ring has been turning relative to its seating – this is confirmed if the seating is also polished.
- If bearing rings have been turning, the mating parts must be either renewed or built up and machined to their original sizes.
- Check the bearing tracks, rollers or balls for surface imperfections.

Renew the needle roller bearings if:

- Rollers, cages or rings are damaged.
- Rollers fall from cage.
- Rollers can be moved noticeably out of alignment.
- The shaft surface is damaged or worn. If this has occurred shaft particles will have entered the bearing, and subsequent early failure is almost inevitable.

If a bearing is judged to be fit for further service it should be carefully washed in clean solvent and immersed in clean oil for temporary storage. If in any doubt whatsoever, renew!

Fitting

Bearings are normally a press-fit on rotating parts and a sliding-fit on static components such as housings. Whenever possible, fit the bearing to the rotating part first. Use a tubular drift or press tool but ensure that it doesn't contact the ring lip since damage may be caused. If a press isn't available when fitting large bearings, it's preferable to either heat the bearing in hot oil or to heat the housing rather than use a large hammer and drift. Always ensure that the bearing abuts squarely against the locating face.

Pressure testing

Ensure that there are no leaks from hoses or pipework before pressure testing is carried out. Readings should be taken using a gauge adequate for pressure testing, either liquid-filled or fitted with a restrictor or damper.

Before adjusting a valve, unscrew the adjusting screw by two or three turns to release any trapped dirt.

Servicing

Service access was another great plus factor on the JCB, especially as regards the engine – there were no fuel or hydraulic tanks around to cause obstructions, and a huge bonnet

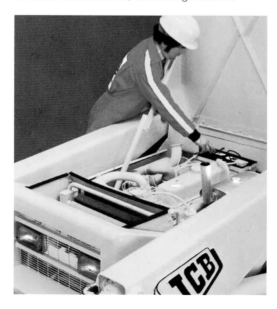

LEFT A large single-piece, rear-hinged bonnet gave all-round engine access to carry out daily checks and maintenance.

gave all-round engine access. Brakes could be serviced without wheel removal and the engine gearbox components could be reached without completely dismantling the machine.

The simple but effective slew mechanism was readily accessible for servicing, there were only 12 pivots to grease on the front loader, and JCB's well-designed hydraulic system meant far fewer components to check – all vitally important considerations, because the harder it was to service a machine the more likely it was to be neglected, which could lead to poor performance and, eventually, breakdowns.

Serviceable parts
- Air filter.
- Hydraulic filter.
- Hydraulic tank strainer.
- Steering filter.
- Fuel filters.
- Oils – see lubricants listing on page 113.

SERVICE SCHEDULE

	Daily	50 hours	100 hours	200 hours	400 hours	800 hours	1,600 hours
BEFORE STARTING ENGINE							
Adjust							
Tightness of wheel nuts	X	X	X	X	X	X	X
Tyre pressures	X	X	X	X	X	X	X
Wheel alignment					X	X	X
Brake adjustment		X	X	X	X	X	X
Clutch adjustment		X	X	X	X	X	X
Fan belt adjustment		X	X	X	X	X	X
Air cleaner hose security		X	X	X	X	X	X
Starter motor, alternator brush gear							X
Change/adjust							
Fuel filter element					X	X	X
Air cleaner element						X	X
Valve clearances					X	X	X
Clean							
Drain fuel sediment trap		X	X	X	X	X	X
Fuel lift pump				X	X	X	X
Clean and test injectors					X	X	X
Air filter vacuator valve				X	X	X	X
Battery terminals			X	X	X	X	X
Grease							
Axle pivot	X	X	X	X	X	X	X
Pump drive		X	X	X	X	X	X
Kingpins	X	X	X	X	X	X	X
All pivot points	X	X	X	X	X	X	X
Kingpost	X	X	X	X	X	X	X
Brake cross-shaft	X	X	X	X	X	X	X
Clutch bearing	X	X	X	X	X	X	X
Coolant pump					X	X	X
Control level ball joints			X	X	X	X	X
Propshaft (Build 5)		X	X	X	X	X	X
Oil all linkages			X	X	X	X	X

	Daily	50 hours	100 hours	200 hours	400 hours	800 hours	1,600 hours
Check levels/leaks							
Axle oil (Build 5)		X	X	X	X	X	X
Slewing actuator oil		X	X	X	X	X	X
Engine oil	X	X	X	X	X	X	X
Engine coolant (anti-freeze)	X	X	X	X	X	X	X
Transmission oil (Manual & Build 3)		X	X	X	X	X	X
Hydraulic oil	X	X	X	X	X	X	X
Fuel system					X	X	X	X
Gearbox oil (Build 5)					X		
AFTER STARTING ENGINE							
Check							
Operation of electrical equipment		X	X	X	X	X	X
Instrument readings	X	X	X	X	X	X	X
Exhaust (excessive smoke)	X	X	X	X	X	X	X
Idling speed					X	X	X
Torque convertor oil level	X	X	X	X	X		
Main relief valve pressure					X	X	X
Auxiliary relief pressure					X	X	X
Steering relief valve pressure					X	X	X
Operation of all services							
Excavator	X	X	X	X	X	X	X
Shovel	X	X	X	X	X	X	X
Transmission	X	X	X	X	X	X	X
Brakes	X	X	X	X	X	X	X
Steering	X	X	X	X	X	X	X
Rams for leaks and rod damage					X	X	X
Maximum governed engine speed					X	X	X
Engine pull-down speed					X	X	X
Torque convertor oil cooler		X	X	X	X	X	X
AFTER STOPPING ENGINE							
Change							
Engine oil				X	X	X	X
Engine oil filter element				X	X	X	X
Hydraulic fluid and clean suction strainers							X
Hydraulic fluid filter element							X
Torque convertor oil						X	X
Gearbox oil						X	X
Axle oil						X	X
Torque convertor oil element					X	X	
Check							
Hoses and pipework			X	X	X	X	X
Tightness of ram cap ends (screwed)			X	X	X	X	X
Engine compression					X	X	X

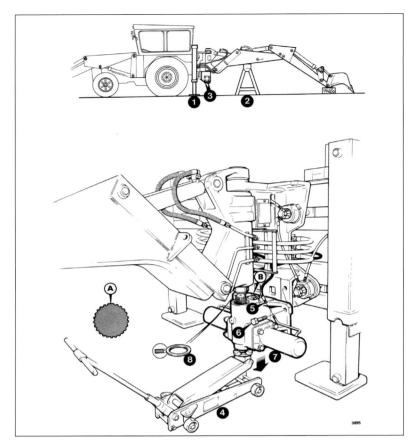

LEFT Removing the slew actuator.

Component removal

Key replacement parts:
- Axle pivot bush.
- Ram eye bushes.
- Hoses.

Body and framework

Slew actuator removal and replacement
The numerical sequence shown on the left is intended as a guide to removal. For replacement, the sequence should be reversed.

When removing
- Position and support the boom and dipper as shown.
- Drain the oil via both plugs at 3.
- Label and blank hoses (5) to ensure correct replacement and prevent the entry of dirt.

When replacing
- Renew the thrust washer (8) with the chamfer facing towards the kingpost.
- Take care to align the master spline (A).
- Refill at B with HD90 oil.

Torque settings

Item	lb/ft	kgf/m
6	600	83

Slew actuator dismantling and assembly
The numerical sequence shown on the illustration is intended as a guide to dismantling. For assembly, the sequence should be reversed.

When dismantling
- To remove items 19, 20 and 21, enter press-tool through the hole for plug 17.
- Note that items 11a to 16a are fitted on later machines in place of 11 to 16.

When assembling
- Position master spline (A) directly above centre tooth B of rack 22.
- Align hole C in bush 19 with grease nipple D (if fitted).

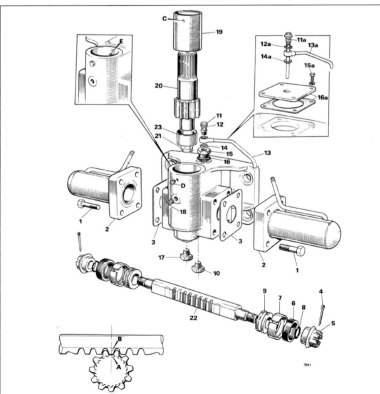

LEFT Dismantling the slewing actuator.

Torque settings

Item	Nm	lb/ft
1	372	275
5*	420	310
10	95	70
17	95	70

* Item 5 may be torqued higher to allow holes for split pin (4) to line up.

Boom and dipper
Removal and replacement
The numerical sequence shown on the illustration is intended as a guide to dismantling. For assembly, the sequence should be reversed.

When removing
- After positioning support 2, stop the engine and operate the boom and dipper controls to vent residual hydraulic pressure.
- Label and blank hoses to ensure the correct replacement and to prevent entry of dirt.

Kingpost/kingpost carriage
Removal and replacement
The numerical sequence shown on the illustration is intended as a guide to dismantling. For assembly, the sequence should be reversed.

When removing
- First remove the boom and dipper.
- Label and blank hoses to ensure correct replacement and prevent entry of dirt.
- To remove pivot pin 13, use a 1in UNF bolt in the tapped hole provided.

Note: Before swinging the kingpost rearwards, lower the hoist so that the top pivot boss clears housing C.

When replacing
- Ensure that when the Hydraclamps are pressurised, clearance is 3mm to 5mm (1/8in to 3/16in).
- Renew thrust washers 12 and 15 with chamfers facing towards the kingpost.

Torque settings

Item	lb/ft	kgf/m
10	600	83

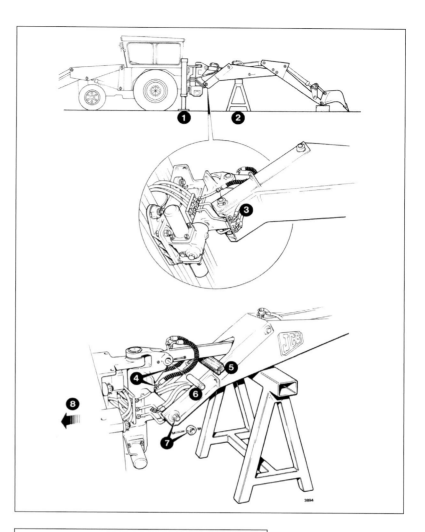

ABOVE Removing the boom and dipper.

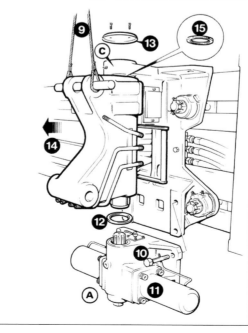

LEFT Removing the kingpost and kingpost carriage.

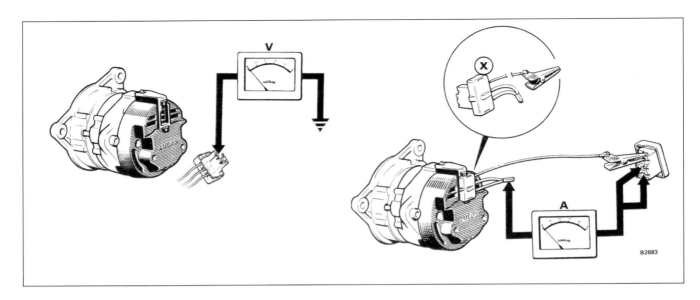

ABOVE Testing the alternator. **BELOW** Hydraulic pressure testing.

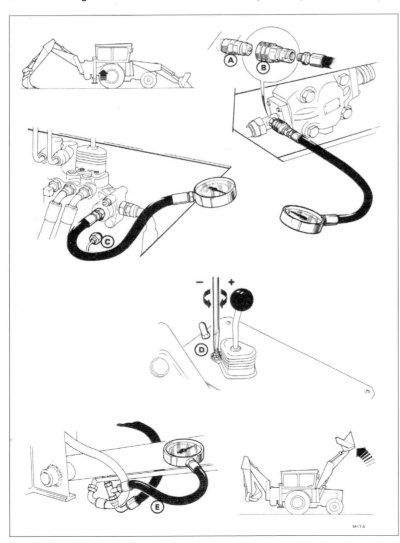

Electrics

Charging circuit check

- Stop engine and turn start switch to 'Off'.
- Withdraw the alternator plug and, using the voltmeter section of a suitable test meter, check for battery voltage between the large terminals and 'earth'.
- If the reading is zero, check the brown cables for continuity, particularly at the starter terminals.

If the voltage is correct, check the alternator output as follows:

- Using a dummy connector, connect the ammeter section of the test meter between the alternator and the alternator plug as shown (positive meter lead to alternator). Note that connections are similar for both Lucas and AC Delco alternators.
- Connect the voltmeter section of the test meter across the battery (observe polarity).
- Start the engine and run at 2,000rpm. As the engine starts, the voltmeter reading should rise rapidly to charging voltage (14V to 15V).
- Switch on all lights and ancillaries and check the ammeter reading. This should be approximately 40A.
- If the ammeter reading is low, check that all charging cables and connectors are cool. A hot connection is 'resistive' and

should be cut back and remade. If the meter reading is low but everything else is satisfactory, stop the engine and carry out further tests.

Troubleshooting

Main relief valve (MRV) pressure testing

- With the shovel and bucket on the ground and the engine stopped, operate all control levers to vent hydraulic pressure from the system.
- Connect a pressure gauge fitted with a snubber to test adapter A. If adapter A isn't fitted, remove plug or connection C and connect gauge into the excavator valve.
- With the engine running at maximum rpm, fully retract a stabiliser ram and note maximum gauge reading, which should equal the pressure specified in Technical Data in Appendix E.
- Repeat the check by operating other excavator services, when a similar gauge reading should be obtained in each case. A low reading on any one service would indicate a fault within that service, *eg* fluid leakage, defective ram seals or defective auxiliary relief valve (ARV). Note that there are no ARVs in the stabiliser services.

If an incorrect reading is obtained on all excavator services, proceed to the shovel circuit as follows:

- With gauge connected at A, run engine at maximum rpm. Fully extend gauge reading.
- Repeat the check with the shovel rams.
- If adapter A isn't fitted, connect a tee-piece into the lift ram feed as at D. Run the engine at maximum rpm, extend the lift rams to fully travel and note the maximum gauge reading. If the shovel circuit pressures are consistently higher than those recorded on the excavator services, the fault is likely to be a leak from between the sections of the shovel valve.
- If all gauge readings are approximately equal, but differ from the specified MRV setting, the MRV (E), which is non-adjustable, must be removed and either cleaned or renewed.

LUBRICANTS

Engine and injection pump	
-10° to 50°C (14° to 122°F)	SAE 15W-40 oil
-15° to 30°C (5° to 86°F)	SAE 10W-30 oil
-20° to 0°C (-4° to 32°F)	SAE 10W oil
Torque convertor/shuttle	
Mobil Delvac 1310 (SAE 10W) or JCB Special Torque oil	
Gearbox/Final drive	
10-speed (to s/n 130301)	HP90 oil
4-speed (Build 3)	HP90 oil
9-speed (synchro)	20W20 oil
Gearbox	
4-speed (Build 5)	20W20 oil
Drive axle (Build 5)	
To machine 138621/5 (dry brakes)	20W20 oil
From machine 138622/5 (oil-immersed)	JCB Special Gear oil
Brake system (Build 5)	
To machine 138621/5 (dry brakes)	Mobil Hydraulic Brake Fluid 550
From machine 138622/5 (oil immersed)	Below -20°C (-4°F) Mobil DTE11
	Above -20°C (-4°F) JCB Special Hydraulic fluid
WARNING: Use of incorrect brake fluid will cause serious damage.	
Slewing actuator	
HD 90 oil.	
Hydraulic system	
JCB Special Hydraulic Fluid, ISO 46.	
Grease points	
Lithium grease.	
Extending dipper	
Mobiltac E or OGL-3 oil.	

Capacities	Litres	US gallons
Engine oil – Manual and Build 3 balanced engines		
steel sump	11	2.9
alloy sump	11	2.8
Engine oil – Build 5	9.1	2.4
Cooling system	13.6	3.6
Torque convertor		
Build 3	20.4	5.4
Build 5	11	2.9
Gearbox		
10-speed	50	13.2
9-speed (synchro)	57	15
Build 5	2.5	—
Drive axle		
Build 5 with dry brakes	25	6.6
Build 5 with oil immersed brakes	29.5	7.8
Fuel tank	68	18
Slew actuator	4.5	1.2
Hydraulic system	195	51.6

Chapter Eight

Marketing

JCB's founder, Joseph Cyril Bamford, understood the value of the brand far earlier than many, and he never tired of finding new ways to promote it. He purchased ten motor-scooters in 1958, on condition that the licence-plate numbers were JCB 1 to 10, and he then transferred the numbers to company and family vehicles.

In 1960, the keyboards on the company's typewriters and telex machines were specially modified to include a key which reproduced the JCB logo.

OPPOSITE 'Another Genuine JCB' sticker was fitted to all 3C Mk IIIs and continued in use until the early 1980s on the replacement 3CX model. This is the author's restored 3C Mk III.

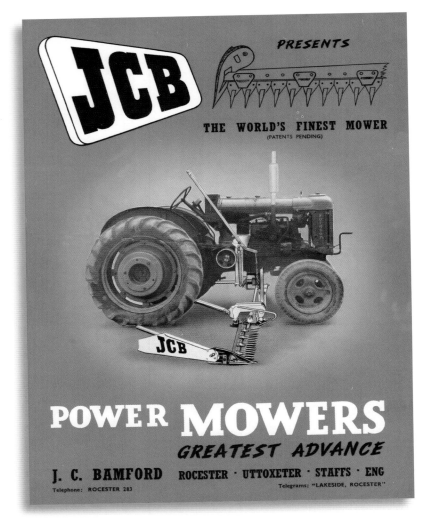

ABOVE The original JCB logo angled up to fit on to the side swath of the mid-mounted mowers in the early 1950s.

BELOW The steel embossed name badges were fitted up until 1980, when they were replaced by vinyl lettering due to cost.

The JCB logo

The JCB logo was produced originally because Mr Bamford wanted to avoid confusion with the old family firm, Bamford Ltd, in nearby Uttoxeter. The logo with its three letters and lozenge-shaped surround was developed by a small design agency in Wolverhampton. The letters were angled up in order to fit neatly on to the slanting side swath of the mid-mounted mowers that Mr Bamford built in the early 1950s.

'When I left the family's agricultural business in Uttoxeter,' Mr Bamford recalled, 'I wanted to preserve, but not trade under, the Bamford name, so I decided to put the emphasis on the initials J and C.'

The letters were pressed into steel nameplates that were welded on to each side of the front loader and on either side of the boom. These steel plates continued in use through the 1960s and 1970s right up until 1980, when they were replaced with black vinyl letters on the last 3C Mk IIIs.

The 1963 version of the logo entailed a 3D effect combined with a drop shadow. This was amended within the publicity department in 1972, and in the early 1980s the shadow was removed.

Advertising

JCB advertised regularly in a variety of construction journals throughout the 1950s, '60s and '70s, one of the most popular being *Contract Journal Magazine*, in which the company placed large, full-colour adverts. Very often these took the form of pull-out supplements with industry-leading colour pictures and very few words. One such campaign was 'Specify JCB', 'Specify JCB for quality and performance', where the message was 'Don't just hire an excavator, specify JCB'. Another campaign, promoting the 3C Mk II, had the strapline 'Most excavators have some (JCB) features, but only a JCB has them *all*!'

The publicity department

The publicity department was set up to assist and support sales by promoting the product, service and company. Four methods were used to make the public aware of JCB's products – advertising, sales, promotional literature,

ABOVE An early advertising mailer posted to customers to promote the JCB 3.

press coverage and sales aids. Numerous different campaigns were used over the years to promote its products, and in the early 1970s a series of advertisements was published on the 'Genuine JCB' theme. JCB was aware that it was becoming a household name for the backhoe loader, just like 'Hoover' had become for the vacuum cleaner, so the problem was that when someone asked a hire company for a JCB they were quite happy when a Ford or Massey was delivered – as far as they were concerned they'd got what they'd asked for. Consequently an advertisement was produced telling people how to identify a *genuine* JCB. Armed with this knowledge, the site foreman actually recognised a Massey and said, 'Take it back, it's not a real one.' In parallel with this advertising campaign a large 'Another Genuine JCB' decal was fitted on the roof of every new machine above the front windscreen.

Hard facts marketing campaign

Market research showed that JCB's advertising wasn't strong enough and it was concluded that, in order to have any impact at all with an advertisement, there were two ways of doing it. One was to advertise a product as brand new and revolutionary, the other was to ensure that if you had something to say then you should say it, and say it hard. So appeared the 'Hard Facts'

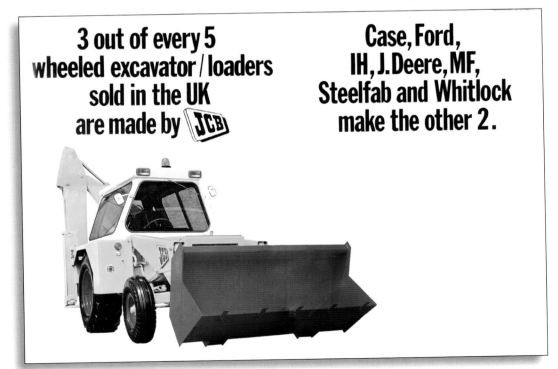

LEFT An advert from *Construction News*, September 1971.

ABOVE 1957 four-page sales leaflet for the Mk I Major Loader. On the front cover it listed the sectors for which the machine was ideally suited.

Sales literature

Literature promoting JCB products dates back to the start, when Joseph Cyril Bamford's wife Marjorie played a large part in the company's development and was involved with the various machinery shows around the country. At one exhibition, Mr JCB had black-and-white leaflets printed that he asked Marjorie to hand out to potential customers. Unknowingly, she presented a leaflet to one of the Whitlock brothers, who were major competitors. They complained to the show organisers that JCB was taking an unfair advantage over other exhibitors, but by that time the JCB stand was full of enthusiastic customers.

New sales literature

Early sales leaflets were basic double-sided sheets that featured a picture of the machine on the front and gave details of its specification on the reverse. This approach continued through the early 1950s, until four-page leaflets were introduced in JCB's corporate yellow and red colours. In the early 1960s the leaflets were changed to incorporate a large picture on the front cover and a full-page picture of the machine in operation on the reverse. Late in the 1960s the leaflets reverted back to a single sheet.

In 1971 the publicity department took a closer look at the company's literature and decided to provide more specification details, by making their leaflets four pages long again and

series of fold-up adverts that were very popular and successful in North America, and that car manufacturers had started to use in the UK (where, for instance, Rover and Ford compared their cars with those made by their competitors).

RIGHT The front cover of a rare JCB range brochure from the mid-1960s, with the 4C alongside the smaller JCB 3. In the background is Wootton Lodge, the 17th century house in Ellastone, Staffordshire, which was Mr JCB's home.

using the front page for machine application shots, the back page for specific sales features, and the two inside pages for specification data. The result was a far more comprehensive and informative leaflet.

Periodic updates had to be made to all sales literature as machine specifications changed, which rendered older versions obsolete. Between the publication of the earliest specification sheet recorded for the Hydraulic Excavator in 1948 to the literature accompanying the final 3C Mk IIIs over 350 different specification sheets and sales brochures were produced for the backhoe range.

Promotional literature was also translated into key languages for those non-English-speaking countries where JCBs were sold. Specification leaflets dating as far back as the JCB 4 in 1960 were printed in Japanese, for example.

Sales training literature

JCB promotional literature was initially designed for its customers to see and use, but the focus subsequently changed to creating literature specifically designed for its own sales team. Just as the service engineer had a service manual, the sales engineer would have a sales manual designed for their reference rather than the customer's. The first backhoe loader sales manual was published for dealers in 1972.

Part numbers literature

A part numbers index system was introduced on printed sales literature from 1957, which listed the abbreviations for the model followed by the date – for example, H/D/1/3/57 was for a Hydra-Digga brochure released on 1 March 1957. The literature number index codes changed in 1960 to a three-digit code, *ie* 401, but the date code was dropped. In 1963 the code format changed again, this time to the prefix 'PD', standing for 'Publicity Department', and a four-digit code. If the brochure was updated and reprinted it would have an edition letter after the part number – for example, PD1046 EDC, the 'C' meaning that the brochure was on its third edition.

With the introduction of the new part numbers index system, which started to appear in 1964, the sales publications and promotional items were allocated the group prefix numbers

ABOVE A 1978 3C Mk III sales brochure.

Sales and service literature for the 3C Mk III		
Part Number	Description	Date
9999/1532	Specification sheet	1977
9999/1627	Extending dipper leaflet	1977
9999/1649	Brochure	1977
9999/1532	Specification sheet	1978
9999/0651	Brochure	1978

9999/, while service publications were preceded by numbers 9801/ to 9809/. The first sales leaflet released in the new format was part number 9999/1001, for the JCB 2.

Walk-around cards

'Walk-around' cards were first introduced and given out at the UK sales conference in 1976, for dealers to remind their salesman of a machine's key features as they walked their customer around it.

ABOVE The 1970s merchandise range included a selection of items for work, leisure and pleasure.

RIGHT The JCB donkey jacket was made from the best-quality Melton fabric with two front pockets and a yellow waterproof shoulder patch screen-printed with the JCB symbol.

Merchandise

From 1975, JCB Service's parts department was responsible for the storage and distribution of all promotional aids, sales literature and stationery. A wide range of literature, promotional aids and clothing was available, including jackets, ties, T-shirts and blue overalls. Due to unprecedented demand for the JCB overalls, coinciding with a decline in world production of cotton, in 1974 JCB was unable to fulfil all its orders. Negotiations therefore started with alternative suppliers in anticipation of a new type of improved style and quality becoming available later that year. A brand new design made from polyester/cotton fabric was issued, this time in bright red with a yellow stripe running down each arm, and corporately identified with sewn-on cloth badges on the left breast pocket and back panel. Specifically designed for machine operators and JCB demonstrators, they were dry-clean only and not suitable for workshop use.

Selection from the JCB merchandise range (1976)	
Description	Price
Dark navy blue tie	£1.90
Red overalls 40in	£8.70
3C Mk III model	£3.31
Donkey jacket	£7.82
Golf umbrella	£2.90

Scale models

The first model of a JCB machine appeared in 1964, in an Airfix construction kit of the JCB 3 backhoe loader. It was available from Woolworths and then from other toy shops, as well as from JCB, priced at £3.

The JCB 3C Mk III was one of the machines around which many successful building and earthmoving businesses were built up over the years. Being a machine that made money for these operators, often in substantial fleets, it isn't at all surprising that many owners and operators developed a definite soft spot for them. For some, this also extended to having models of these iconic machines, both in their offices and in their homes. Models were also bought by collectors who appreciated what these machines could do, but weren't old enough – or simply

RIGHT The JCB 3 Airfix model was introduced in 1964, priced at £3.

CENTRE The Corgi JCB 3C model was available in both white and yellow cab. The model was copied from a real 3C owned by J.C. Balls Ltd in Derbyshire.

didn't have the opportunity – to own and operate full-size machines. For others, securing a model was a reminder of a visit to the JCB factory. What's more, this applies just as much now as it did in the heyday of 3C Mk III production.

It isn't surprising that over the years the 3C Mk III has been reproduced in model form a number of times. Some of these models date from the same production period as the full-size machines, whilst others have become available in more recent years. All differ in their scale, level of detail and availability.

Some models started life at JCB's Rocester factory, being built as styling models, often

BELOW In-house model-makers produced handmade models of all JCB's new machine concepts to review the styling and make early changes before prototype machines were built. Mr Bamford with son Anthony Bamford in the drawing office in 1978 with a model 3C Mk III.

ending up on display in the factory. Made at quite large scales, their construction was often of aluminium, plywood and plastic.

Diecast models

What the collector could buy at the time was a JCB 3C Mk II produced by NZG, a West German manufacturer of diecast models. These well-made diecast models were relatively expensive in their day and were used both as toys and as promotional tools. There were two basic 1:36 scale models to collect, both with yellow cabs; these were the loader only and backhoe loader versions. Their shade of paintwork varied somewhat over the years, as the shade on some models was noticeably more orange than others. Other details that varied included the work lights, chevrons on the stabilisers and hydraulic filter decal on the right-hand mudguard.

Many variants of the 3C Mk II model were produced, and, again, the most obvious difference between them was their yellow paint shade. The darker yellow models had either 'high-profile' front tyres, as used on the yellow-cab 3C Mk II, or 'low profile'. The heavily tinted cab-glass variant resembled the real 'Caribbean cab' machine which was exported to hot

TOP 3C Mk II yellow cab model made by NZG in Germany, now very sought-after amongst collectors. A loader-only version was also produced.

ABOVE The 3C Mk III was also produced by NZG in Germany. It has the taller cab and beacon on the roof, although these are susceptible to being knocked off and are often missing.

RIGHT A limited number of gold 3C Mk II models were produced and given out by JCB to VIP customers.

ABOVE The 'digger in a bottle' glass sculpture was very prestigious and only given to VIP visitors and to a lucky few employees as retirement gifts for long-term service.

climates. A special edition gold model was also made of the 3C Mk II, which was given only to VIP visitors and as a retirement gift to long-serving staff.

NZG also produced a 3C Mk III model, its most noticeable difference from the 3C Mk II model being the taller cab with a roof-mounted beacon on the later model and the yellow decal on the right-hand mudguard.

Almost 40 years later these long-since obsolete models are still traded amongst collectors, and have clearly at least held their value, with well-looked-after boxed examples changing hands for sometimes substantial sums. Others still grace the collections of the original purchasers, and won't be parted with at any price!

Sometimes special models also appeared. One of these was the 'digger in a bottle', commissioned by JCB and made by the Lichfield Glass Sculpture glass-blowing company, which specialised in putting sculptures into bottles. The digger is about the size of an NZG one, made in glass, and in a wine bottle. One of the sculptured glass backhoe loaders in a bottle was presented to

LEFT A press launch was held in the theatre at JCB's headquarters to launch the Britains 3C Mk III model in 2013. It was unveiled on the stage alongside the full-size 3C Mk III which had just been restored by the author.

ABOVE These two rare JCB 3 miniatures were made by Kent Vintage Models in the 1990s. That on the right has the later cab with 'up and over' opening rear window.

ABOVE Tomica-Dandy produced an excellent model of the JCB 3D, which is very rare. There are two variants: one with red decals and the other with black.

HRH Prince Charles during his visit to JCB in November 1977.

In more recent years a wider selection of 3C Mk III models have become available. The most recent one is sold under the Britains brand. Like almost all toys, and also many lower-priced collectors' models, this is mass-produced in the Far East and sells to children and collectors alike.

For the collectors, a market has already opened up to 'correct' details that are technically incorrect with the model, such as by offering rib-type front tyres instead of the traction tyres that would have been an option at the time. Additional beacons and, more particularly, a range of buckets are also offered by several specialist suppliers in the UK. Other versions are offered that have been weathered to make them look more at home amongst the dirt and mud in dioramas.

Moving up the price and detail scale, Martyn Henley of Kent Vintage Models produced several 3C Mk III variants over the years. Made in the UK from a mixture of white metal, polyurethane resin and plastic, three models were originally available in the mid-1990s, but are now very rare as they were only in production for a short time.

RIGHT 'Laying the Clay' model produced by Border Fine Arts, part of a limited edition of just 1,750 models. When introduced in 2002 it cost £550.

The rarest is the larger-scale JCB 3, which had more detail in the cab, steerable wheels, moving stabilisers legs, movable loader and backhoe and hydraulic hoses.

Border Fine Arts, based in Scotland, produced a collection of eight JCB cast models that were handmade and hand-painted. The collection was unique, with its focus being on nostalgic scenes, and the prestigious models were issued as limited editions. Each model scene was given a name, such as 'Laying the clay' which featured a JCB 1 digging a trench for drainage with clay pipes on the side ready to fit.

We next come to the models from Classic-Combines. Hand-built in the UK in very small numbers and available only on a built-to-order basis, these are all to a rather larger scale of 1:16. The larger scale is important, as it enables these models to be more accurately described as 'collectors' engineering models'. Made from a mixture of precision-cut sheet brass and cast brass, on final assembly each one is bolted together just like the original, and with proper nuts and bolts. The only difference is that being as small as 1mm over the threads, these brass hexagon-headed nuts and bolts are considerably smaller than the originals.

This level of engineering detail also extends to the pivot pins and bushes on the functioning loader and backhoe elements, which are cross-drilled and bolted, just like they should be. Even the diagonal pattern on the cab floor plate, the foot pedals and the safety stickers are present inside the cab. Whilst the exact number clearly varies with each model and its precise build specification, typically there are around 1,000 separate parts and assemblies in each model.

All three basic versions of the 3C Mk III have now been catered for, with manual, Powertrain and Brockhouse torque convertor-type transmissions now tooled. Complementing these are a wide range of optional features such as alternative tyres and a host of attachments, including 6-in-1 clam shovel, fork options, dozer blades, pole planters and hand-held hammers to the front, plus extending dippers and a selection of buckets, rippers and hammers to the rear. Even the famous JCB Jaw Bucket has now been reproduced, along with its associated backfill blade and tarmac cutter. All these buckets feature miniature plough bolts to hold on the replaceable bucket teeth. All in all, these are the closest thing that anyone can buy to the large-scale models that are dotted around the JCB factory.

ABOVE An exact scale model of the author's 3C Mk III, made by Classic Combines (www.classic-combines.com), which features the Rock Breaker on the side of the machine and Jaw Bucket. The model is fully movable and was handmade from over 1,000 parts.

Epilogue

CELEBRATING 70 YEARS OF INNOVATION

ABOVE The 3CX Platinum machine was introduced in 2015 to celebrate JCB's 70th Anniversary. Seen here standing side by side with the 3C Mk III, the paint scheme is modelled on the colours of the 3C Mk III machine owned by the author. It makes it very easy to see the shared ancestry of the two JCBs.

On 23 October 2015 JCB marked its 70th Anniversary. Today JCB has 22 factories, 11 in the UK and the others in India, the USA, Brazil and China, employing more than 12,000 people.

This special milestone has been marked by the production of a limited edition version of the iconic 3CX backhoe loader that fuelled the company's success. A total of 70 of the

RIGHT In 2013 JCB celebrated 60 years of building backhoes and the completion of 1,000,000 machines.

RIGHT The 3CX anniversary machine comes with a comprehensive customer gift pack that includes a pair of retro-style red overalls, retro polo shirt, a set of spanners and a model of the 3C Mk III next to a 3CX, all in a fantastic presentation case. It even comes with a carpet for the cab!

LEFT The limited edition 3CX anniversary machine has its unique number printed on the authenticity plate on the front windscreen.

special machines will be made in the striking livery last seen 40 years ago on the JCB 3C Mk III model in 1980. These limited platinum edition backhoes, which are built at JCB's world headquarters in Rocester, Staffordshire, come complete with red buckets, full white cab and red wheels instead of the customary black and yellow finish.

Fitted with 109hp JCB Tier 4 Final EcoMAX engines manufactured at JCB's plant in Derbyshire, they have a top speed of 40kph (25mph). They're also fitted with JCB's LiveLink telematics system, which enables owners to monitor the whereabouts of their machine and to manage its service needs.

In addition they have a host of features to showcase the machine's attributes, including six-speed autoshift transmission, TorqueLock air-conditioned cab with heated air-suspension seat, coffee machine, quick-hitch mounted 6-in-1 clam shovel with ram guard protection, hydraulic excavator quick hitch, rear light guards and streetpads.

Those customers lucky enough to have one of these machines are presented with a special operators' pack that contains limited edition items such as a set of spanners in a bag, as originally supplied. The pack also includes a pair of overalls that are a replica of the 1970s red overalls.

Since JCB's first machine was manufactured in 1953 the company has produced more than 600,000 backhoes and now sells them in 120 countries. JCB has also been the world's biggest manufacturer of backhoes for 15 years in succession, with a range that spans 40 models from the compact 1CX through to the mighty 5CX.

The specification of the new 3CX is certainly impressive compared to the old 3C Mk III, as can be seen from the table:

JCB MK III AND 3CX COMPARISON

	3C Mk III (1978)	3CX Contractor (2015)
Engine manufacturer	British Leyland 4.98	JCB EcoMAX
Power	52kW (70hp)	81kW (109hp)
Engine capacity	3.8 litres	4.4 litres
Engine torque	258Nm	516Nm
Aspiration	Naturally	Turbocharged
Operating weight	7,910kg	8,271kg
Transmission	2WD – 4-speed	2WD/4WD – 6-speed
Road speed (Powertrain model)	25kph (15.5mph) 27.2kph (16.9mph)	40kph (25mph)
Pump type	Gear	Piston pump
Hydraulic pressure	138 bar	251 bar
Hydraulic flow	126lpm	165lpm
Excavator type	Fixed	Extending
Shovel capacity	0.8m	1.0m
Loader lift capacity at full height	3,099kg	3,499kg
Shovel tear-out force	4,026kgf	6,170kgf
Fuel tank capacity	68 litres	160 litres
Width	2.43m	2.35m
Length – travel position	6.36m	5.91m
Height – travel clearance	3.31m	3.48m

Appendix A
Pricing

UK pricing, 1978 (VAT not included)

- Fitted with a 4-cylinder 52kW (70hp) diesel engine with torque convertor transmission with four forward and four reverse gears.
- Standard tyres front and rear.
- General-purpose front shovel, 2,020mm (80in) wide, 1.04m^3 (1⅜yd^3) capacity.
- Full weather protection cab, soundproofed.
- Interior light, headlights, sidelights, working lights, rear light reflector indicators.
- Fully instrumented control panel with remote ignition and oil warning lights.
- Full power steering.
- Front and rear windscreen washers, windscreen wipers, heater/demister, cigarette lighter, sliding offside window.
- Hydraulic kingpost clamping.
- One 650mm (24in) backhoe bucket complete with teeth and side cutters.

UK retail price

Standard transmission model with nine forward and three reverse gears	£12,505
Torque Convertor transmission with four forward and four reverse gears	£12,876

Factory-fitted options

Safety cab	£280
Arctic start engine heater including pre-set timer	£260
Face-level fan	£17
Hazard warning system	£22
Cab soundproofing	£135
Engine soundproofing	£186
Steering-wheel knob	£4.50
Vandal-proof instrument cover	£43
Heavy-duty battery upgrade	£4
Supplementary tool kit	£105
Suspension seat (semi-damped)	£61
Lap seat belt	£6.50
Exhaust extension	£5
Exhaust cover	£3
Tarpaulin cab cover	£46
Rubber street pads for stabiliser feet	£65
Excavator bucket mounting bracket (on shovel)	£21
Fire extinguisher	£22
Tooth guard	£21
6-in-1 multipurpose clam shovel in lieu of standard, complete with oversized tyres	£1,400
JCB Roadbreaker	
31.7kg single hammer**	£638
31.7kg double hammer**	£2,760
Extending dipper including 18.4/15 x 26 12ply heavy-duty tyres	£1,075
Extending dipper**	£655

Buckets

305mm (12in) backhoe bucket complete with teeth and side cutters	£192
400mm (16in) ditto	£149
500mm (18in) ditto	£152
650mm (24in) ditto	£172
800mm (30in) ditto	£191
950mm (36in)	£280
1,524mm (60in) ditch maintenance bucket	£205
Backfill blade, 1,524mm (60in) (not for 500mm/18in Jaw Bucket)	£210
Ripper tooth	£205
Asphalt cutter	£190
Bucket extension 305mm (12in) – pair to fit ditching bucket	£121
Forklift attachment – arm-mounted	£385
Forklift attachment – shovel-mounted	£236
Poleplanter – single (complete with 'V' steady frame)	£1,687
Jaw Bucket* 305mm (12in)**	£810
500mm (18in)**	£890
650mm (24in)**	£935
Backfill blade for Jaw Bucket	£182
Front shovel teeth including nuts and bolts (per set)	£48
18.4/15-26, 12-ply rear tyres – oversize heavy duty industrial***	£245

* Price inclusive of pipework and valve block etc, factory fitted.
** To comply with tyre-loading regulations, 18.4/15-26, 12 ply tyres are mandatory.
*** Where oversized tyres are fitted to the front wheels only, the loader dig depth will be reduced slightly. Price given assumes standard tyres.

Appendix B
3C development

This appendix documents the significant developments that were made to the 3C model, culminating in the Mk III model that was launched in February 1977.

Chassis and bodywork
Engine canopy/bonnet
- Ventilation port incorporated to improve engine cooling.
- Introduced from machine serial number 112986.
- Construction changed to a cross-ridged pressing without diagonal bracing, which made it easier to lift and improved its appearance.

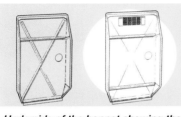

Underside of the bonnet showing the additional diagonal bracings

Piston ring
- Due to a change in the engine design, Leyland 4-98 engines were built without the bottom piston ring. This improvement was designed to prolong engine life, although higher oil consumption during the running-in period was expected. Introduced in December 1978.

Front grille
- 60% grille coverage was introduced, which meant that the grille could take 8kph (5mph) impact without serious damage. Introduced from machine serial number 109506.
- Grille restyled to give improved protection to the hydraulic pump and to improve appearance. Introduced from machine serial number 112986.
- A headlamp guard option was introduced in January 1977.

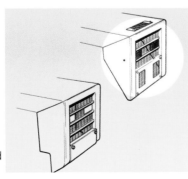

Front grille

Hose clamping
- Stud that retained the clamp on the rear wheel cover plates changed to a bolt, which eliminated the possibility of hose chafing. Introduced in July 1978.

Baffles
- Anti-recirculation baffles were added between the radiator and the grille to improve cooling. Introduced in mid-1974.

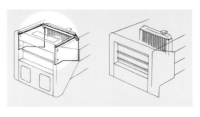

Addition of baffles to improve cooling

Loader
6-in-1 clam shovel
- A reinforcing plate was added inside the clam base to improve performance from stiffer clam. Introduced from machine serial number 10949.

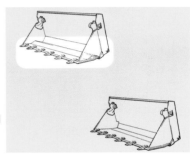

6-in-1 clam shovel

Forklift attachment
- Changed to forged steel, which is simpler to manufacture and had a longer life. Introduced in October 1973.

Slew mechanism
- Rack and pinion teeth were shot-peened during manufacture to give longer life, as surface conditioning inhibits fatigue cracks. Introduced from machine serial number 109709, June 1975.

Slew locking pin
- A second locking bar added to improve security in travelling position. Introduced from September 1975.

Improved slew cylinders
- Slew cylinder changed to one-piece extrusion using the Biginelli process (for forming bottomed tubular members from metal blanks). This ensured maximum strength, uniform bore thickness and high reliability. Introduced in December 1972.
- 127mm (5in) slew cylinders introduced to give 23% increase in slew torque. Introduced on extending dipper machines from machine serial number 114303, and all 3C Mk IIIs.
- Diameter of lower slew bearing increased to 114mm

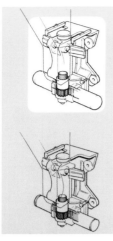

Slew mechanism

(4½in), which increased bearing area by 6% for longer life. All 3C Mk IIIs.
- Actuator mounting bolt reduced in diameter to make it easier to set to correct torque. Introduced from machine serial number 115683.

Extending dipper
- Introduced to the UK and Europe in April 1976 and export markets in May 1977.
- Locking pin and locating holes added to give greater safety when travelling.

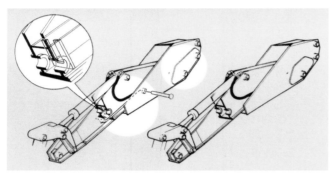

Extending dipper

- Reinforced plates added around extension ram rod-end pivots and enlarged to encompass dipper ram pivot to add extra strength to give longer life in hard digging conditions.
- Extension side plates. Introduced in September 1975.
- Locking pin March 1976.
- Dipper ram plates April 1976.

Excavator buckets
- New teeth with 60% more material and recessed bolt heads. This extended wear life, made them easier to replace and didn't score the surface of excavation. Introduced in September 1975.
- New curved profile for 3C gave better filling and dumping characteristics. Introduced on 3C Mk III machines.
- Back plate of 800mm and 950mm (32in and 37in) buckets thickened near pivot pins to provide greater resistance to fatigue in heavy digging conditions. Introduced in two widths, 460mm (18in) and 600mm (24in) versions, introduced in the UK in December 1974 and for export regions in February 1976.

Jaw Bucket
- The ram was shortened by 12mm (½in) in August 1976 to improve clearance when the jaw was fully open.
- A protection strip was added around the pivot at the same time, to give better protection of the pivot ends during hard digging.
- Also at the same time an access hole was added to provide simplified lubrication of the ram pivot.

Hydraulics
Kingpost clamping
- Improved check valve design to give greater resistance to failure. Introduced from machine serial number 110503.
- New seal type introduced to reduce leakage and extend life. Introduced from machine serial number 110054.
- The cylinders were hard chromed and the surface finish was improved to reduce wear and improve sealing. Introduced in October 1975.

Hydraulic pump
- A redesigned gear pump delivered increased hydraulic flow of 125 litres (32 gallons) per minute. Introduced in July 1970.

Extending dipper
- In September 1975 the extending dipper evolved to be powered by its own, separate hydraulic service and was strengthened. The dipper could be extended or retracted at full speed whether operated simultaneously with another excavator service or not. The revised circuit included an additional pump, mounted in tandem with the main systems pump.
- At the same time an additional, replaceable wear strip was added to the slide of the extending dipper to allow for in-service wear.

Double pump
- The double pump manufactured by Commercial was replaced in June 1978 by the Hamworthy pump, which went on to be replaced by a Hamworthy double pump with a one-piece body.

Hydraulic fluid level gauge
- A clearly marked white background was introduced to make it easier to check the fluid level. Introduced in April 1978.

Hydraulic filter
- Changed to water-resistant impregnated

Hydraulic filter

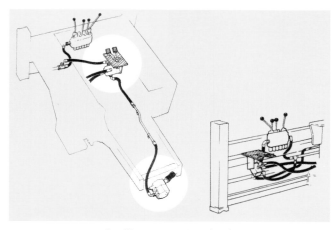
Auxiliary excavator circuit

paper to provide better protection against any water suspended in the hydraulic fluid. Introduced in March 1976.

Auxiliary excavator circuit
- Powered from a separate pump to provide simultaneous operation with excavator and better control. Introduced from machine serial number 114625.

Neutral circuit
- Pump type changed to give equivalent performance at lower replacement cost. Introduced from machine serial number 100612.
- Snap-on pressure check point added to give rapid check without loss of oil. Introduced from machine serial number 113980.

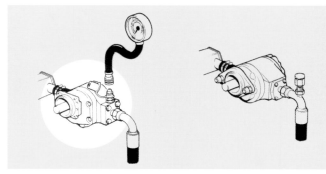
Neutral circuit

- Pressure check point turned through 90° to improve access. Introduced in August 1976.

Loader circuit
- A three-spool valve block was introduced as an option, providing automatic self-levelling of the shovel, two-speed dump and float position on arms, plus auxiliary hydraulic service for loader. Introduced on machine serial number 104067 in May 1973.

Engine and transmission
- New cerametallic clutch and bonded brake linings. Introduced in March 1972.

Manual gearbox
- A new synchromesh gearbox was introduced for manual transmission machines, consisting of two sections: a three forward and one reverse speed full synchromesh gearbox, followed by a three-speed constant mesh range change. Combined, they provide nine forward and three reverse speeds. Incorporated improved bearing capacities, increased face widths on gears, improved mainshaft nut fixture and a heavier duty mainshaft, which made it easier to drive with better control and increased reliability. Introduced from machine serial number 130302 in August 1978.

Air cleaner
- A new air cleaner with secondary safety element had increased capacity and lower pressure drop, which offered improved engine protection, called for less frequent renewal and improved engine breathing. The safety element fitted inside the standard element and provided protection to the engine in the event of the standard element becoming holed. Also acted as a safety element, as it remained in position when the standard element was removed, preventing loose dust from falling into the inlet manifold. Introduced from machine serial number 129343 in June 1978.

Exhaust system
- Exhaust silencer was sprayed in aluminium to give longer life and improve appearance. Introduced in July 1976.
- On applications where free-flow material could spill out of the shovel on to the bonnet, a plate became available in August 1978 to fit around the exhaust pipe. This prevented material from entering the engine compartment.

Sump guard
- Extended beyond the engine to provide greater protection to the pipework underneath. Introduced in July 1974.

Tachograph cable
- A heavier-duty, nylon-lined version was introduced to improve reliability. Introduced in June 1976.

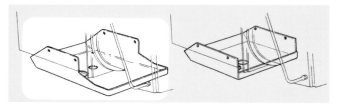
Sump guard

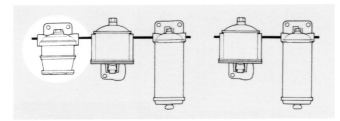
Fuel system

Fuel system
- A sedimenter was added into the fuel line before the lift pump to improve fuel cleaning, provide better protection for the engine and improve reliability. Introduced from machine serial number 110553 in June 1965.

Torque convertor reaction member clutch
- Sprag-type replaced roller-type to give longer life. Introduced in May 1974.

Spur to hub reduction
- Reduction on torque convertor machines changed from 5.333:1 to 4.4182:1 (except in Germany and Austria). This increased the maximum travel speed by 27%. Introduced in February 1974.

Brakes (torque convertor builds)
- Improved braking efficiency achieved by the introduction of larger diameter transmission brakes known as 7 x 4, referring respectively to the external and internal diameters (in inches) of the friction discs. Comparable dimensions of previous brakes were 6.5in and 3.5in. Lining area increased by 10%, which provided a 60% increase in braking efficiency for greater safety, manoeuvrability and output.
- Linkage modified to accommodate the bigger brakes. Introduced from machine serial number 129503/3 in June 1978.

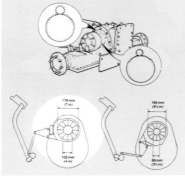
Brakes

Introduction of oil-immersed brakes
- Further improvement in braking efficiency was achieved by introduction of oil-immersed brakes with full hydraulic operation. Each brake comprised five friction plates and six reaction plates, all running in the axle oil which was increased in capacity to 29.5 litres (6.5gal) and had to be JCB special gear oil. The brakes became fully self-adjusting, not requiring the operator to compensate for wear. The units were externally mounted for ease of servicing and maintenance. The oil-immersed brakes weren't interchangeable with dry brakes due to half-axle casings being incorporated. Introduced from machine serial number 138622/5 in March 1980.

Oil-immersed brakes – Powertrain only

Steering and axle
- A new heavy-duty cast beam with transverse steering ram, wider centre pivot and longer king pins was introduced. This provided greater strength, longer bearing life, better ram protection, and improved weight distribution and travelling stability.

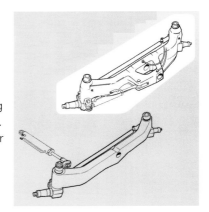
Front axle

Front wheel hubs
- Double lip seal introduced to reduce the loss of grease, which meant that the hubs no longer required routine lubrication, therefore reducing maintenance times. Introduced from machine serial number 126080.

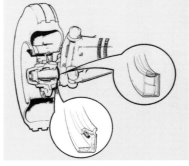
Front wheel hubs

Hydraulics
- Pipework modified to suit new axle. Introduced from machine serial number 68151.
- Snap-on pressure check point added to facilitate a rapid

check of pressure without loss of oil. Introduced from machine serial number 113980.
- Dust cover added to check point to improve protection against the ingress of dirt. Introduced in November 1976.
- Strainer added to ensure longer component life, as oil was kept cleaner. Introduced from machine serial number 110759.
- Manual control unit introduced to give the same performance at lower cost. Introduced from machine serial number 116516.

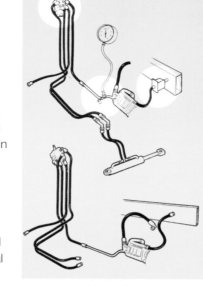

Hydraulics

Cab
- A hinged door was introduced to allow walk-in access, which was easier and safer for the operator. Introduced on the 3C Mk III in February 1977.

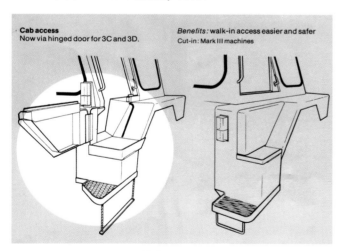

Cab access

- Vandal-proof instrument panel covers were introduced in November 1974.
- A fire extinguisher option was introduced in September 1976.
- A seat belt was introduced in June 1972 on the 3C Mk II range. It was optional in the UK.
- With the introduction of the 3C Mk III, a 150mm (6in) higher cab was introduced to provide more space and improved visibility from 14.2% increase in cab volume and 27% increase in glass area.
- Sliding right-hand windows introduced to improve ventilation and provide better visibility in humid conditions. Introduced from machine serial number 65501, January 1973.
- A laminated windscreen was introduced to provide greater safety for the operator. Introduced in December 1973.
- Cab soundproofing fitted as standard, reducing cab noise by 5dBA (UK and Eire).

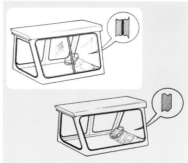

Cab size/glazing

Throttle
- High-specification throttle cable introduced sealed with a rubber boot at all exit points. Packed with low temperature grease, it improved the hand throttle control and repositioned stronger safety sprint to injection pump. This provided higher reliability in cold and dusty environments, easier operation and full travel of the fuel control lever. Introduced from machine serial number 129446 in June 1978.
- Two types of throttle were fitted as alternatives from machine serial number 129445. Each type had a different-length cable to suit skid build or JCB Powertrain machines.

Instrument panel/steering wheel/ windscreen washer
- New layout panel made of rigid urethane foam. Introduced from machine serial number 68151.
- Stop and cold-start controls were moved to a separate panel, giving improved cable runs which were sealed, with low-temperature greasing to provide more durable, improved styling and instruments and controls that were easier to use. Introduced from machine serial number 129446 in June 1978.
- The windscreen washer was made a standard offering to provide improved basic specification. The 'Kigass' washer system with push pump operation had its water piping concealed in the cab frame pillars. Introduced on 3C Mk III.
- The steering wheel was restyled and raised by 100mm (4in) to improve appearance and driving position. Restyled steering wheel introduced in October 1977. Raised steering wheel introduced in 3C Mk III machines.

- The removable rear window was supported on a fixed rod track with positive locking to provide smoother operation, improved sealing and vandal-proofing. Introduced on 3C JCB Mk III in February 1977.
- The wipers were changed to pantograph-type and the rear wiper was mounted on a removable frame to provide greater visibility from a 16% increase in cleaned area, with no need to park the wiper before removing the window. Introduced on 3C Mk III machines.

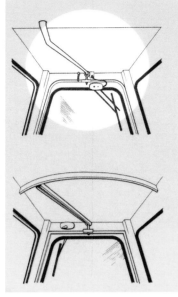

Rear window/wiper

1979: One of the first products to be built at JCB's own transmission factory in Wrexham, North Wales, was the JCB Powertrain transmission fitted to the last of the 3C Mk IIIs

Roadbreaker
- A hydraulic divider was fitted with an additional relief valve to provide automatic 50:50 split of flow without adjustment, which gave even better protection against over-pressurisation. Introduced from machine serial number 108990 in June 1975.
- Roadbreaker changed to include twin hoses in place of coaxial type and 'live' hose reel. This improved stowage and paying-out, and made it easier to fit other tools. Introduced from machine serial number 131101/3 in October 1978.

JCB Powertrain models

New drive line

The introduction of a new drive line for the 3C Mk III was JCB's answer to the market's demand for faster, easier travelling, both on and off road, and faster cycle times on loading and site-stripping work. At the introduction of the Powertrain – or 'Build 5', as it was also referred to – the opportunity was taken to review other aspects of the 3C Mk III. The most significant changes took place with the engine compartment and the operator's cab, but the Powertrain's new vertical exhaust stack, which made the exhaust system more efficient, was its most obvious outward distinguishing feature.

The new JCB-designed and manufactured Powertrain transmission comprised an individual transmission and axle

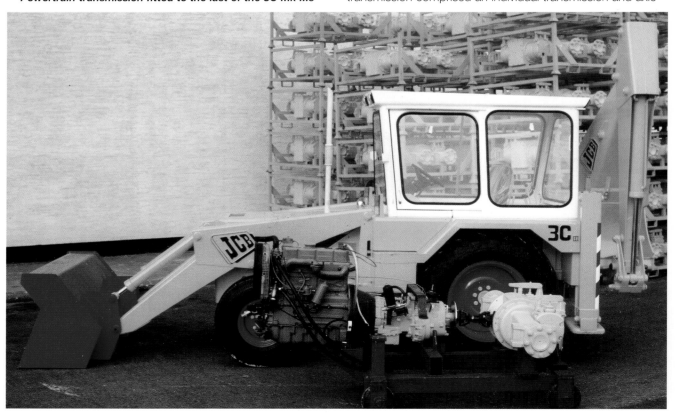

JCB Powertrain

and used a Hardy Spicer driveshaft to take power from the gearbox to the axle. The shaft was the sliding type, which reduced axle loads into other transmission components and made installation easier.

Introduced in June 1979 (serial no 134191/5), the first two machines were built for JCB's own demonstration fleet, to take to construction shows and use for the company's dancing diggers routine, which had by now become a popular attraction. The new transmission was used on the last 3C Mk IIIs ahead of the introduction of the new 3CX in 1980.

The Powertrain models were deemed to be the best 3C Mk IIIs, as they provided faster and easier operation, both on-site and on the road. By using an individual transmission and axle, the best component match was achieved. Unlike the Brockhouse torque convertor, which was slow to take up drive when engaged, the new version responded 'on demand' to variations in engine speed, and low-inertia clutch packs ensured rapid directional change by the reversing shuttle.

Synchromesh gears

The provision of a 'power disconnect' pedal allowed the operator to declutch the reversing shuttle drive completely, while full synchromesh sliding hubs and helically-cut constant mesh gears on the main gearbox enabled gear changing on the move, which couldn't be done on the previous Leyland transmission. This meant faster travel between sites, cutting down on non-productive road time and eliminated the need to stop the machine before changing gear, which was particularly frustrating when approaching hill climbs. Also, when maximum engine speed was called for at the loader hydraulics during loading, declutching the transmission left the engine free to supply full power to the pump and improved loading performance.

Transmission testing

The drive axle was an all-JCB-built component, based on principles tried and proved on a whole generation of backhoe loaders, but with uprated structure and mechanics. One axle-case laboratory rig was reported to have tested at up to four times the working load and withstood more than 7.5 million cycles before the research team gave up trying to break it!

Axle testing

Axles fitted to prototype machines successfully clocked up 25,000 hours before permission was given to build them into production machines. Taper roller bearings at all locations contributed to high dynamic strength and double metal-faced seals kept lubricant and dirt in their respective places.

Upgraded brakes

The new brakes were an upgrade from the previous mechanically operated brakes on the Leyland skid unit transmission. The new hydraulic operation had twin hydraulic master cylinders which were compensated to ensure equal braking effort at each wheel when both pedals were in operation, regardless of any difference in disc wear which may have been brought about by single pedal (skid turn) usage. Hydraulic operation, in combination with the large heat dissipating area of 7 x 4 discs, improved braking power, minimised brake fade and reduced brake adjustment maintenance.

Parking brake

The parking brake was bolted to the gearbox output flange, which meant that the mechanical advantage of the axle was between it and the wheels, providing extra holding power on steep gradients. Separating the parking brake from the main brake meant it provided an emergency system should it ever have been needed. The parking brake disc diameter was 279mm (11in).

The brake system operated via a cam-type lever applying pressure to the brake calliper, with operator effort being transmitted from the driving cab by a heavy-duty cable. Used in conjunction with the standard overcentre lever, the system gave an overall multiplication of 22 times, sufficient to hold the machine on a one-in-three slope.

Engine mounting

The engine was now resiliently mounted to the machine mainframe using rubber mountings, reducing the vibration transmitted to the operator and making their working environment more comfortable, while also eliminating the need for an internal balancer inside the engine. Other changes inside the engine included high-duty, larger-diameter gudgeon pins and four (instead of five) piston rings.

Changes under the bonnet

A number of engine ancillaries were also moved around to improve routine service access. The air cleaner was now bolted to the mainframe instead of the battery tray, giving better access for filter renewal.

The alternator was moved to the left-hand side of the engine to reduce load on the alternator bearing (slack side of drive belt), and to avoid overheating by the exhaust manifold.

The engine oil dipstick was made simpler to use, being moved away from the obstruction of the exhaust silencer.

Control panel

A distinctive redesigned control panel housed the brake fluid reservoir and handbrake lever as well as the normal auxiliary controls and warning lights. The handbrake lever was the over-centre type already well proven on the JCB loading shovels and telescopic handler.

The inside of the fenders provides a flat surface for the operator to store his belongings or as a seat for work colleagues

Instrumentation comprised combined engine speed and hourmeter and a fuel gauge; switches and key start were mounted on an angled face to prevent the ingress of water. All engine and transmission functions (alternator charge, oil pressure and temperature, coolant temperature) were monitored by both visual and audible warnings, giving the operator an immediate indication of a fault condition, so that quick action could be taken to prevent further damage. The audible warning was wired into the 'Handbrake' and 'Air filter blocked' warnings.

Standard equipment

Standard equipment included hazard warning, screen wash and automatic stalk switching for direction indicators, horn and headlight dip/flash. A two-speed cab heater was built into the control panel bulkhead and the engine stop and cold-start controls were fitted to their own sub-panel.

With operator comfort in mind, a cigar lighter and 12V power point for the JCB optional kettle were provided.

Driving controls

The lowered line of the new transmission allowed for a completely level driving cab floor, which gave the JCB design team freedom to position the main driving controls to suit the needs of the operator. A direct-acting gearshift lever, combined with the 'declutching' action of the power disconnect pedal, made gear changes fast and positive. The shuttle-reversing lever was mounted, as before, on the left-hand side of the control panel.

Loading was made faster and easier, with the heel and toe convenience of the brake and power disconnect pedals enabling the operator to hold the machine on the footbrake, declutch the transmission and, with the right foot, accelerate the engine to give maximum power to the loader hydraulics.

By using individual components, the best combination of matching units had been put together to reduce operator fatigue and to keep man and machine performing at maximum efficiency throughout their shift. Separate units also meant easier servicing.

Several features identified the new cab. A restyled control panel, along with a revised driving control layout, were made possible by the flat floor that replaced the lowered drive line. This design made driving the machine easier, leaving the operator free to concentrate on the job in hand.

Torque convertor

Because a torque convertor generally multiplied engine power in proportion to the speed difference between the engine and the transmission, it provided maximum turning effort when it was most needed, *ie* when starting away from standstill or accelerating a fully loaded machine. Absence of direct mechanical coupling (power is transmitted only through the energy of moving oil) meant that engine effort was taken up smoothly without shock, reducing wear and tear on transmission components.

A temperature and oil pressure gauge for the torque convertor (Brockhouse models only) is fitted on the far left of the front console next to the ignition barrel and bank of switches for the lights and cab heater

Appendix C
Attachments

The development of the backhoe loader over the decades has resulted in it being regarded as one of the most versatile types of machine on a construction site. This is one of the many factors that has contributed towards the success of the JCB backhoe loader.

During the 1950s and 1960s the number of attachments offered was fairly limited and consisted mainly of a general-purpose bucket or clam shovel on the front loader and buckets of different widths on the excavator end. But, as the concept evolved and became more established, JCB focussed its attention on promoting the machine in different sectors within the market and developing more attachments that enabled it to work on a wider number of applications. The attachments shown in this section were developed over the years to take full advantage of this adaptability.

The concept of the backhoe loader was very popular in the house-building and general construction sectors, but JCB wanted to expand its popularity into other markets,

A selection of excavator attachments

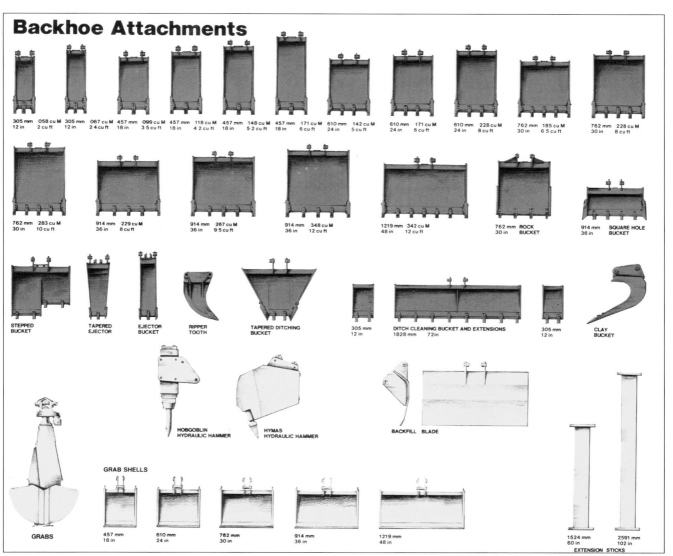

including agriculture, industrial, water and drainage projects and public utilities. To enable the machine to be sold into this wide variety of sectors they had to invest in developing it to fit and power a number of new attachments.

Excavator buckets and attachments

The type of buckets used on the JCB 3C Mk III were first introduced in March 1971. They were recognisably different in shape to the previous design and had a radial back which increased tear-out by 30%. New-style teeth were made from alloy steel and were extra long with integrated side cutters, which bolted on to the toeplate from inside the bucket. In addition to the profile change, the 'speed' and 'power' pivot points were transferred from the bucket to the tipping link.

The radial back to the bucket made it easier to dig trenches and increased tear-out force compared to previous bucket designs

SAE bucket ratings

The Society of Automotive Engineers (SAE) is an American organisation that recommends engineering standards and practices. SAE standards are the basis for a number of regulations laid down by the Federal State governments and they're recognised worldwide, particularly as regards earthmoving machinery.

The SAE's calculations, examples of which are mentioned in the following sub-section, are based on the physical dimensions of the bucket only, without the addition of side cutters or other accessories and without regard to bucket action.

General-purpose buckets

All standard excavator buckets were of similar welded steel construction, with a tough, hardened toeplate and curved profile for clean and efficient digging. They had bolt-on teeth and the two outside teeth incorporated side cutters for increased strength and long life. They were available in a range of sizes from 350mm (14in) wide to 950mm (37.5in). Capacities varied from 0.06m³ (2ft³) SAE to 0.3m³ (10ft³) SAE. Buckets were, with a few exceptions, completely interchangeable with the range of JCB backhoe loaders, but because of differences in hydraulic power and excavator geometry the bucket performance varied with the machine to which it was fitted. For this reason, buckets were designated 'standard profile' or 'deep profile'.

General-purpose excavating bucket with bolt-on teeth and side cutters

The bucket crowd linkage had two positions, enabling the bucket to be operated for maximum power (185° rotation) or maximum speed (201° rotation) of heaped loads and 90° trench bottoms. A heavy-duty design of bucket tooth was fitted and held in place by a plough-bolt that was flush-fitting to improve trench finish. The profile shape of the teeth meant that they self-sharpened, which maintained high penetration performance.

All were of welded steel plate of varying thickness. The back plate and side plates were manufactured from 6mm (0.24in), pivot plates from 10mm (0.39in), front side plates from 12.5mm (0.49in) and toeplate from 20mm (0.79in).

The toeplate was protected by hardness-weld to Brinell 575. Side cutters and teeth, and annealed and hardened.

The JCB Jaw Bucket had more success with local authorities as it was particularly useful for lifting broken tarmacadam and removing kerb stones

Bucket type	Size*		Capacity SAE				Weight		No of teeth
	mm	(in)	rated		struck		kg	(lb)	
			m³	(ft³)	m³	(ft³)			
Standard profile	305	12	0.07	2.5	0.06	2	96	212	3
profile	400	16	0.09	3	0.07	3	103	227	3
	500	20	0.13	4.5	0.10	3.5	110	243	3
	650	26	0.18	6.5	0.14	4.5	134	295	4
	800	32	0.24	8	0.17	6	156	344	5
	950	37	0.30	10	0.20	7	192	423	5
Deep profile	500	20	0.18	6	0.14	5	115	254	3
	650	26	0.24	9	0.19	7	145	320	4
High tear-out	350	14	0.06	2	0.05	2	93	205	3
	500	20	0.11	4	0.09	3	106	234	3
	650	26	0.16	5.5	0.12	4	128	282	4
	800	32	0.20	7	0.15	5.5	152	335	5
	950	37	0.24	9	0.18	6.5	160	353	5

* Including side cutters. For width less side cutters, deduct 51mm (2in).

Jaw Bucket

The idea of the Jaw Bucket was originally designed, patented and developed by Wain-Roy Inc of the USA, from whom JCB secured sole selling rights throughout the world except for North America.

First introduced on the 3C Mk II in the UK in December 1974 and for export in February 1976, it was initially available in two widths, 460mm (18in) and 600mm (24in). Eighteen Jaw Buckets were given to UK customers and a few to overseas customers for evaluation and feedback. Their most significant reaction prior to this was that the cost was fairly high when compared with a conventional excavator bucket. As with the JCB Roadbreakers, it was therefore necessary to sell by demonstration rather than description.

The JCB jaw was a special, heavy-duty excavator bucket, able to perform all the normal digging functions as well as having the ability to handle large pieces of material such as tree roots, rocks or masonry. It could also be used to lift lengths of timber, lintels and other awkward items.

Due to the additional weight over a standard bucket and its ability to lift heavy objects, the Jaw Bucket required fitment of the stronger 12-ply, 18.4/15-26 tyres to comply with tyre loading regulations.

Of the three available sizes of Jaw Bucket, the 600mm (24in) was the most popular. The one seen here is digging up asphalt during a pipelaying job

Operation of the Jaw Bucket

The bucket consisted of a two-part hydraulically operated clam, profiled as a bucket. The jaw ram was operated by two pedal controls adjacent to the operator's right foot. The left pedal opened the jaw and the right pedal closed it. Back-up pressure at the ram provided continuous clamping action, which is what enabled the bucket to be used for gripping materials such as rocks, tree-roots and piping.

Hard-wearing parts

With more rugged construction than conventional buckets, a built-in guard protected the hydraulic cylinder from damage. The toeplate was hard-faced to Brinell 575 and the side cutters and teeth were hardened to resist wear, and specially heat-treated so they didn't become brittle.

The back of the Jaw Bucket was fully enclosed to protect the hydraulic ram and hoses from damage. The hydraulic clamping force was over 2,087kgf (4,600lbf)

The JCB Jaw Bucket was ideal for removing tree stumps and handling awkward material

The Jaw Bucket was heavily marketed during the 1970s, particularly to local authorities for road maintenance work, as it was a useful attachment for pulling out kerbs and loading into lorries, but it was an expensive option at the time.

In April 1977, in response to popular demand, a new 305mm (12in) jaw was introduced to enable narrow trenching work to be carried out. A major application was the laying and removal of kerbstones, where the narrow jaw kept ground disturbance to a minimum.

3C Mk II fitted with the JCB Jaw Bucket and twin Roadbreakers

To further increase the jaw's versatility two quick-fit attachments were designed for the 600mm (24in) Jaw Bucket, which could be easily picked up and held in position by its clamping action. A tarmacadam cutting wheel was introduced in August 1976 for semi-solid surfaces. A backfill blade for trench filling had already been introduced by this date. These items could be mounted and detached in a matter of seconds, and were so useful as to be considered essential tools by local authorities.

Repair work to main services often required the digging of a trench in a tarmacadam surface. Here the jaw and auxiliaries could be used in four ways to speed up the work. By fitting the cutting wheel, the road surface could be precisely marked out for the required trench work without affecting the surrounding area; the rotary cutting disc could be rolled on soft road surfaces to make cuts up to 50mm (2in) deep in one pass.

The 'clam' action of the jaw was then used to lift out the tarmac in pieces, before the bucket was used to excavate the bulk of the material in the normal way until it reached the proximity of the service pipe or cable. To avoid the possibility of damage, the pipe or cable could then be carefully exposed by using the smooth-edged back blade of the open jaw.

The tarmacadam cutting wheel was held in position on the Jaw Bucket. Rolled on soft road surfaces the rotary disc could make cuts up to 50mm (2in) deep

For the building industry, the JCB Jaw had many advantages. Initial site clearance on smaller sites often involved the removal of short lengths of hedging and the filling-in of the associated ditch, while in previously occupied areas the derelict properties had to be demolished and the old foundations excavated.

A backfill blade could also be easily picked up by the Jaw Bucket for efficient back-filling of trenches and levelling of ground

Square-hole bucket

A special 960mm (38in) bucket, with a rated SAE capacity of $0.10m^3$ ($3.5ft^3$), was available and profiled for digging square holes for manholes, poles, foundations for stations, pylons and piles etc. Square-hole digging was achieved by a combination of increased bucket rotation and the relationship of the bucket width to the working radius.

The bucket was fitted with a special pivot linkage to increase bucket rotation and had extra-high tear-out up to 35% greater than normal excavating buckets. Its all-welded steel construction had double-thick cutting edges on the side plates. The bolt-on teeth and side cutters were optional.

Square-hole bucket

Ditch-cleaning bucket

This shallow-profiled bucket designed for waterway silt and weed removal had a functional shape featuring a flat bottom and rounded back, providing smooth, regular shapes. It was self-draining due to its perforated sides and bottom drainage holes. The pivot plates incorporated spring steel bushes and grease nipples for lubrication. Available in

Ditch-cleaning bucket

The extending dipper gave additional reach over the side, which was ideal for ditch maintenance work

1,525mm (60in) and 1,830mm (72in) widths, a 2,135mm (84in) version could also be achieved by attaching extension pieces.

Tapered ditching bucket

The tapered ditching bucket, or trapezoidal bucket as they were also known, was designed for forming clean, taper-sided ditches, particularly for land drainage applications, and was popular with farmers. The tapered sides were formed to a predetermined angle that helped to prevent the side walls from collapsing, and the flat bottom facilitated water drainage. Available in two sizes, the 30° tapered ditching bucket had a width of 381–1,676mm (15–66in) while the narrower 60° bucket was 308–1,067mm (12–42in) and could be supplied either with or without side cutters.

The tapered ditching bucket was used on farms for digging drainage ditches. Its tapered sides helped to prevent the side walls from collapsing

Tapered ejector bucket

The tapered ejector bucket was designed for forming clean, taper-sided ditches for the excavation of mud, clay and other materials that would stick inside a normal bucket. The taper-shaped bucket featured a hydraulically operated internal flap that forcibly ejected the material. The taper of the bucket also created a flat-bottomed V-shape trench requiring the minimum amount of rubble fill in land drain construction. Although it was a highly suitable attachment for building and agricultural contractors working in service trenching and drainage work, and particularly useful for land drainage applications, its sales were limited.

Trenching was performed by the retraction of the boom and dipper and not by bucket rotation, as the bucket was secured to the dipper by two link bars, enabling the bucket ram to operate the ejector flap. The bucket could also be used in the conventional way by removing the dipper ties and reconnecting the tipping link to the outer pivot point. The flap then formed the back plate of a deep-profile bucket. The total weight was 150kg (331lb).

The bucket was available in 305mm (12in), 298mm (11.75in) and 203mm (8in) widths. One tooth and two side cutters were fitted as standard.

Tapered ejector bucket

3C Mk III replacing railway sleepers in North America in 1979. It has been specially modified to enable it to drive on rails and has adapted rear-frame legs to provide additional sideways stability

Clay spade

The clay spade was a special-purpose tool for handling sticky materials that would be impossible to work using a standard bucket. It was used for excavation of difficult-to-dump, non-free-flowing materials such as heavy clay. The curve of the reinforced spade was designed in such a way that the material, as it was being cut, would be forced away from the spade surface so that it dumped cleanly when the spade was inverted. Its slim shape, with a slightly tapered profile with pointed end, enabled easy penetration into the ground and was non-stick due to the absence of side plates, which allowed the material to be easily discharged. With an overall width of 273mm (11in), it was available with two optional teeth and weighed 94kg (207lb).

The clay spade was designed without side plates, to prevent material sticking to it

Ripper tooth

The ripper tooth was designed for breaking up hard, compacted surfaces such as tarmacadam and concrete, and was used for road maintenance and demolition applications. The short distance from the point of attachment at the dipper tip gave the ripper tooth a particularly high tear-out force, whilst its curved profile enabled the tip to slide under the object that needed to be moved. To maximise its power of penetration, it utilised the boom, dipper and bucket rams to pry out material. Its alloy steel-hardened tip featured a quick-release locking pin which could be driven in or out from either side for easy replacement. Its cast body with integral reinforcing webs provided great strength, and its curved shape gave unrestricted radial cutting action when rotated. Overall width was 48mm (1⅞in) and it weighed 74kg (163lb).

The ripper tooth was ideal for breaking up tarmacadam and concrete and was popular for road maintenance and demolition applications

Backfill blade

The backfill blade was a tool for refilling trenches with the material previously dug out, providing fast backfilling of trenches, particularly in locations with limited access. It could also empty and clean out square-cornered trucks. It was strong enough to stand up to light-duty scraping and levelling work. Its angled plate featured a fully welded integral mounting plate. It had a width of 1,524mm (5ft), a height of 711mm (2ft 4in) and weighed 171kg (377lb) including pins.

Backfill blade

Powerbreaker

The dipper-mounted JCB Powerbreaker was the only hammer available that was fully matched to the performance of the 3C Mk III. Its applications included breaking through rock when trenching, road-breaking and road maintenance, breaking slag out of furnaces and demolition of building foundations and concrete bases.

The hammer weighed 230kg (507lb), which was light in comparison to more modern breakers, and gave 45kgf/m (325lbf/ft) per blow at up to 800 blows per minute. In comparison to the average hand-held pneumatic hammer, which was widely popular at the time, the JCB Powerbreaker provided ten times more blows, so was a great deal more productive.

Operation

Powered by its own hydraulic circuit, it was available as a factory option or could be fitted to existing machines or as an adaptation to the JCB jaw installation. The Powerbreaker was operated from within the cab by a simple foot-pedal control. This allowed the operator to keep both hands free to operate the normal excavator controls and consequently maintain full control of the machine at all times. Compared with hand-held breakers, the additional benefit was that the operator was kept well away from the actual point of breaking and therefore protected from any potential hazard, as well as keeping warm and dry!

Being a hydraulic hammer, the noise levels were much lower than with pneumatic hammers, and for that reason it was more acceptable in urban environments. The unit mounted on to the standard bucket pivot pins and could remain fitted on the dipper, as it was designed to fold up within the boom and dipper when stowed for transportation purposes. The installation included a warning light and buzzer for hydraulic oil temperature and the high-pressure hydraulic oil filter.

Hose guides were fitted to the side of the tipping link to keep the feed and return hoses from splaying out and getting damaged. A selection of moil points, flat chisels and spades for tarmac and pile-driving applications were available.

The JCB Powerbreaker was a popular option with local councils for road-breaking and road maintenance work. It provided ten times more blows than the average hand-held pneumatic hammer at the time

Loader attachments

Standard equipment for the front loader arms was a general-purpose, flat-bottomed shovel with a parallel top spill guard. It had two functions: to perform a variety of front loader duties including digging, loading, stripping, levelling and carrying; and to provide a stabilising platform

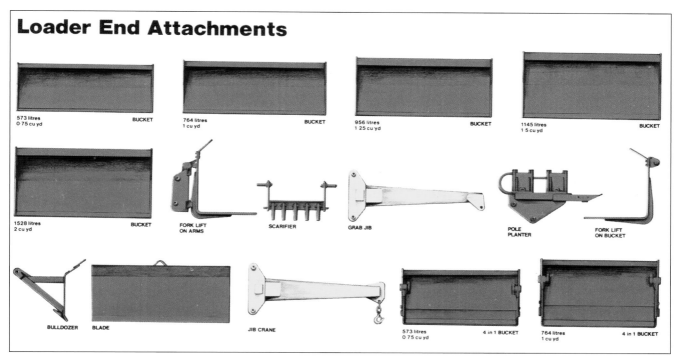

Range of attachments for the front-end loader

when the machine was excavating. It could be positioned flat bottom on the ground or rotated forward for two-point contact, providing greater anchorage to prevent the machine pulling backwards when excavating in hard material. It was constructed from 10mm (³⁄₈in) steel plate and was reinforced at points of maximum stress. Bolt-on teeth and tooth guards were available as an option.

Four different shovel profiles were available to suit a variety of applications – general-purpose, curved, straight-sided (for high-volume loading), and heavy duty, which featured a spill guard.

Flip-over forks

This attachment adapted the machine for handling palletised materials without taking away its bulk-handling facility. It was designed for applications involving both forklift work and normal shovel work, *eg* building sites, store yards, foundries, timber and scrap yards etc. It was ideal for handling logs, pipes, pallets and scrap as well as useful for moving plant and materials around site.

The two forged forks were mounted on a crossbar attached to the sides and top of the shovel. They could be adjusted laterally by pushing them across to accommodate the size of the load. Two rests mounted on the back of the shovel enabled the forks to be swung back for normal loader applications.

Available on both the general-purpose shovel and the 6-in-1 clam shovel, the tines were 38mm (1.5in) thick forgings that could be adjusted to any position along the crossbar. The forks were 914mm (36in) long, 101mm (4in) wide and weighed 157kg (346lb). Their safe working load was at 610mm (24in) centres.

Shovel-mounted forks

Arm-mounted forklift

A heavy-duty forklift attachment was also available, which replaced the front shovel and fitted directly on to the loader arm pivot pins. Designed for a wide variety of material-handling applications, it was employed on building sites for unloading, stacking and re-handling palletised bricks and pipes etc. The tines were infinitely adjustable across the frame to suit the pallet size or load being carried.

The all-welded box and channel-section

The heavy-duty forklift attachment fits in place of the shovel and has the advantage of better stability and load visibility

steel frame carried two forged forks on a crossbar attached to the sides and top of the frame and incorporated mounting and ram pivots and top guides for the two tines. The forks were 914mm (36in) long, 102mm (4in) wide, weighed 157kg (346lb) and had 38mm (1.5in) width tines.

Jib crane

Another attachment that fitted in place of the loading shovel, this converted the machine into a one-tonne capacity crane and was a useful attachment on construction sites and yards for lifting materials around site as well as for loading and unloading lorries. It consisted of a welded-steel pressing 'T'-piece provided with forged steel shackle and hook. Maximum lift height was 2.97m (9ft 9in) and its safe working load was 1,016kg (2,240lb).

The jib crane could lift over 1,000kg to a height of almost 3m (10ft) and was often used for carrying roof trusses on house-building sites

Pole planter

As far back as 1967 a hydraulic rotating lifting fork was developed for the JCB 2, 2B and 3C models for the handling and planting 'H'-type power poles. Equipped with a rear-mounted steady frame, two different versions were available for carrying and erecting single or double poles.

Fitted in place of the excavator front shovel, the unit could be attached in ten minutes. The additional hydraulic service required was taken from a three-spool valve block as used with the 6-in-1 clam shovel. Flexibility of the attachment also made it suitable for planting and transplanting trees and for the erection of tubular steel street-lighting columns.

The unit consisted of a loader arm-mounted platform and a 'V' steady frame mounted at the back of the machine. The platform comprised a base attached to the arms and a pivoted section that could be rotated by a hydraulic ram. The pivoted section carried clamping chains for the pole and two forklift-type tines. The maximum safe working load for the single-pole planter was 1,016kg (2,235lb) and it could manage a maximum pole length of 12.2m

Visibility to the pole was excellent, enabling the operator to accurately plant a pole in less than 15 minutes

(40ft). The double-pole planter consisted of a loader arm-mounted platform on which were mounted two hydraulically operated clamps and safety chains. Adjustable stops provided choice of pole spacing, and hydraulic clamps firmly secured the poles in carrying and erecting positions.

The method of operation was to load the pole on to the carrying frame by using the two forklift-type tines and secure it with the chains. It was then lifted by the loader arms and rotated through a maximum of 100° until it was parallel with the machine. In this way, the pole could be safely carried and distanced above the excavator in the horizontal position. When the hole was dug by the excavator or auger, the machine then approached in first gear and manoeuvred so that the pole was presented to the hole by lowering the loader arms and adjusting the angle by use of the bucket ram. After partial manual backfilling the pole was trued-up using a plumb line and a theodolite, sited at the next location. The pole was then released and final backfilling and tamping was carried out with the excavator bucket. The average time taken for excavating, planting and backfilling was ten to fifteen minutes.

In view of the extra clamping rams, this attachment required an auxiliary hydraulic service to operate the pole clamps, whilst the standard loader hydraulics were used to manoeuvre the poles into position. A three-spool valve block with extra lever and pipework was required to replace the standard two-spool valve block if fitted.

6-in-1 clam shovel

The JCB 6-in-1 was a single front-loader attachment that converted the machine from a normal loader to a highly versatile unit with many different applications in earthmoving and materials handling. Optional for front general-purpose shovels, it was the most popular of all loader-end attachments. The hydraulically operated clam could be quickly adjusted to perform a number of operations, including digging, bulldozing, loading, grading, grab work and back blading.

A high-specification 3C Mk II with JCB Jaw Bucket on the rear and hand-held Rock Breaker mounted to the side, site stripping on the land opposite the JCB factory, which is now the JCB golf range

Popularised in the USA muckshifting industry on wheeled and crawler loaders, the clam shovel subsequently became popular in the UK and elsewhere, particularly for building and civil engineering projects. Closely following the expiration of the patents for the Drott '4 in 1' shovel, JCB released its own version with enhanced specification in 1971, which was known as the JCB 6-in-1.

Its typical applications were many, including hard digging and bulldozing, stockpiling and loading with normal or bottom dumping, fully controlled selective grading and scraping, rubbish-tip control and landfilling, scrap breaking and handling, grab loading, pipe laying, log handling, backfilling footings, trenches and fences etc, demolition work and tree and hedge grubbing.

Performance

The shovel capacity was 0.76m^3 (1yd^3) with a width of 2.03m (6ft 8in), which was just slightly wider than the machine. This meant that the wheels were inside the width of the shovel to enable the machine to dig below the ground for basements etc.

Large 92mm (3½in) bore rams provided a tremendous clamping force of 5,270kgf (11,617lb), which could open up to 840mm (33in) to clamp large objects to a maximum weight of 2,890kg (6,374lb) and lift to a maximum height of 3.16m (10ft 4.5in). An excellent discharge angle of 30° meant it could dump material into high-sided vehicles up to 2.69m (8ft 10in) tall.

Design

The 6-in-1 clam shovel was constructed in two parts: a back section attached to the loader arms and a front clam hinged to the back at its top. It was actuated by two hydraulic rams behind the back plate that, like the main shovel, were designed and built by JCB. These were mounted out of harm's way on the reverse of the back blade. A full-length reinforcing strip behind the back plate provided great strength to this area of the shovel.

Serrated edges on the clam sides provided extra grip for grabbing timber cuttings. Large castings were welded on to the clam shovel pivots, and the pivot pins were EN19C induction-hardened to a depth of 2mm (0.08in) to reduce wear, as was the one-piece toeplate, which was parallel with the bottom plate for accurate site levelling. Six shovel teeth were available as an option, which provided greater penetration when working in hard material. The shovel back blade was heavily reinforced by the side plates and a reinforced lower plate for extra rigidity. A tooth guard was an optional extra that provided safety when travelling on the highway.

Cutting angle and level indicators

Following the same basic shape as the standard general-purpose shovel, the 6-in-1 included a flat top that enabled the operator to judge the angle of the cutting edge. As a guide to the depth of cut when grading, an indicator strip on the top right-hand side of the back blade and a mark on the clam pivot casting showed the opened jaw's height from the ground. The figures on this indicator were in inches and the measurement was taken from the bottom of the cutting edge as it rested on the ground, to the bottom edge of the clam.

Because of the need for an additional hydraulic circuit to control the clam action, the 6-in-1 was usually factory-fitted, but a machine so equipped could also have a standard shovel or other attachments fitted when required.

Bulldozer blade

The bulldozer blade equipped the machine for general bulldozing applications, including levelling of undulating ground and backfilling of trench runs etc. This wasn't a popular attachment. Operators preferred the 6-in-1 as they didn't have to attach and remove it.

Constructed from channel section, frames were welded to each side of the blade. These carried the pivots and locking pins used to attach it to special lugs welded to the shovel. These were supplied loose with the unit, for fitment by the customer or dealer. Two 'V'-shaped prongs were welded to the bottom rear of the blade that located with the toeplate of the shovel to provide a firm, reinforced seating. The toeplate was angled at 30°, which was the optimum angle for efficient dozing. An angled top provided a spill guard for maximum load retention. A lifting eye was positioned in the centre of the top of the blade to facilitate easy attachment and removal.

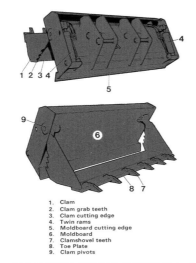

The key components of the 6-in-1 shovel

1. Clam
2. Clam grab teeth
3. Clam cutting edge
4. Twin rams
5. Moldboard cutting edge
6. Moldboard
7. Clamshovel teeth
8. Toe Plate
9. Clam pivots

The bulldozer blade fitted on to the front shovel but was cumbersome to fit and remove and was not a popular option. Scarifier shown fitted to underside of shovel

Appendix D
Machine identification and serial numbers

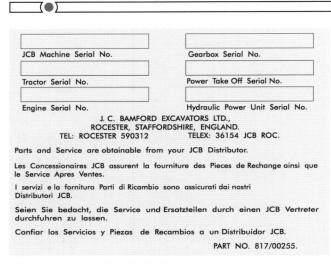

The data plate of the 3C Mk III contains the machine's unique serial number that can be traced to the day it was built

All 3C Mk IIIs are identified by a unique serial number that is stamped on to an aluminium data plate approximately 18cm x 13cm (7in x 5in) in size riveted to the nearside of the machine on the loader tower. The following information explains the various serial numbers to be found on 3C Mk III data plates.

With the introduction of the BLMC Skid in 1964–65, all machines fitted with the skid unit were marked with a letter 'N' (for Nuffield) after the machine serial number. This continued until approximately 1968, when JCB deleted the 'N'.

The serial number of all 3C Mk III machines consisted of six digits, allocated in a single sequence across the entire production period (February 1977 to October 1980). This sequence started at 125000. Those machines fitted with a torque convertor had an additional code of '/3' (for 'Build 3') while '/5' signified the later Build 5 type that was fitted with Powertrain transmission.

The serial number was designated on all plates in the following format:

```
                3CIII 130059/3
Machine type ——|        |——Sequential serial number
```

Note: On Mk I, Mk II and Mk III machines fitted with manual transmissions there was no oblique stroke after the machine serial number. On machines fitted with Leyland skid units rather than Powertrain, the tractor serial number had the prefix SK1/.

With the introduction of the JCB-build torque-convertor skid units, a new serial number was stamped on the skid. Details of the serial numbers are given in the next column:

JCB-build torque-convertor serial number
Stamped on engine/radiator mounting pad.
    (a)   (b)
eg SKO   000000
(a) Denotes skid unit build type: SK1 denotes standard build; SK2 denotes Germany/Austria.
(b) Denotes skid unit serial number.

Engine serial number
Stamped on a pad at the rear of the inlet manifold.
    (a)   (b)   (c)
eg 4/98NT 0000   000000
(a) Denotes engine type – 4 = number of cylinders; 98 = bore in mm; NT = Non-turbo.
(b) Engine build list number.
(c) Engine serial number.

Transmission serial number
Stamped on top of final drive casing at rear.
    (a)   (b)
eg 10B   000000
(a) Denotes gearbox type.
(b) Gearbox serial number.

Torque convertor serial number
Stamped on plate fixed to convertor casing.
    (a)   (b)   (c)
eg 24111 00000 A0
(a) Convertor build list number.
(b) Convertor serial number.
(c) Build date code.

Production of the Mk III range commenced at serial number 125000, which was built on 2 February 1977 for the drawing office at JCB. The first batch of 40 customers' machines were built over the ensuing few weeks and exported to France. The table below shows the serial number breaks that identify the year of manufacture:

Year	Serial number
1977	125000
1978	127534
1979	132152*
1980	137499

* The JCB Powertrain transmission ('Build 5') was introduced in 1979 from serial no 134191/5, built on 26 June.

The last 3C Mk III built was serial number 141351/5. Production ceased in October 1980.

Appendix E
Specifications

3C MkIII SPECIFICATION

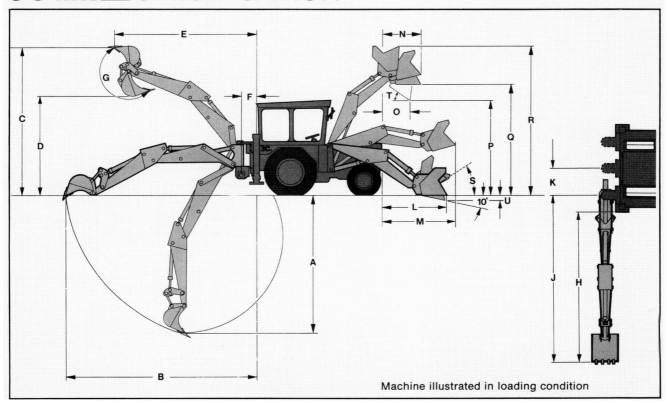

Machine illustrated in loading condition

Excavator performance

Maximum digging depth	4.12m (13ft 6in)
Maximum rear at ground level from stabiliser centre	5.49m (18ft 0in)
Maximum height	4.90m (16ft 1in)
Maximum load-over height	3.35m (11ft 0in)
Maximum reach at full height	3.58m (11ft 9in)
Stabiliser centre to kingpost centre	0.47m (1ft 6½in)
Bucket rotation:	
Power	185°
Speed	201°
Maximum side reach from outside stabiliser feet	4.32m (14ft 1in)
Maximum side reach from kingpost centre	4.95m (16ft 3in)
Kingpost travel	0.59m (1ft 11in)

Loader performance

Reach at ground level	1.78m (5ft 10in)
Maximum reach	2.03m (6ft 8in)
Maximum reach at full height	0.90m (2ft 11in)
Reach at full height, dumped	0.60m (1ft 11¾in)
Maximum dump height	2.96m (9ft 8½in)
Maximum load-over height	3.33m (10ft 11in)
Maximum working height	4.53m (14ft 10½in)
Roll back at ground level	30°
Dump angle	27°
Maximum dig depth, bucket inclined 10°	0.15m (6in)
Shovel tear-out force	4,026kgf (8,876lbf)
Cycle time to full height, shovel loaded	5.6 seconds
Cycle time down, shovel empty	4.4 seconds
Cycle time dump	3.2 seconds
Lifting capacity at full height	3,099kg (6,832lb)
Loader arm tear-out force	4,651kgf (10,253lb)
Forks (shovel mounted) safe working load	1,016kg (2,240lb)
Arm-mounted forklift	1,016 kg (2,240lb)

Extending dipper

Maximum digging depth, normal	4.24m (13ft 11in)
Maximum digging depth, extended	5.21m (17ft 1in)
Reach at ground level, normal	5.56m (18ft 3in)
Reach at ground level, extended	6.48m (21ft 3in)
Maximum load-over height, normal	3.66m (12ft 0in)
Maximum load-over height, extended	4.12m (13ft 6in)
Dipper tear-out force, extended	1,327kgf (2,925lbf)
Lifting capacity of bucket pivot at max extended rear*	544kg (1,200lb)
Lifting capacity of bucket pivot at max reach (no bucket fitted)*	990kg (2,184lb)
Lifting capacity of bucket pivot at max reach (no bucket fitted)*	544kg (1,200lb)
Dipper tear-out force	1,982kgf (4,370lb)
Bucket tear-out force (power position):	
Standard profile bucket	4,695kgf (10,350lbf)
Deep profile bucket	4,052kgf (8,932lbf)
High tear-out	5,269kgf (11,615lbf)
Cycle time:	
Boom lower from full height	2.5 seconds
Bucket closed to open	2.25 seconds
Bucket open to closed	3 seconds
Swing 180°	3 seconds
Crowd in	3 seconds
Crowd out	3 seconds

* The safe working load depends upon which excavator bucket or attachment is fitted. It's calculated by subtracting the weight of the bucket or attachment from the stated lift capacity. All lift capacities quoted are at maximum reach. For example, the safe working load of a standard 350mm bucket is calculated as follows:
990kg (lift capacity of machine) – 96kg (weight of bucket)
= 894kg safe working load.

Weight

Standard fully operational with 650mm (26in) excavator bucket	7,093kg (15,633lb)
Torque convertor	7,174kg (15,813lb)
Weight distribution, excavator offset, slewed and locked, no excavator bucket or operator, including standard tyres:	
Front axle	2,005kg (4,418lb)
Rear axle	4,893kg (10,783lb)

Transmission

Manual synchromesh

Clutch	Single dry plate type, 325mm (12.8in) diameter, 19,610mm (30in^2) friction area.
Gears	Five forward and one reverse with high/low ratio selection on each.

Maximum speed (standard tyres):

	Low		High	
	kph	mph	kph	mph
1st	2.8	1.8	3.6	2.2
2nd	4.5	2.8	5.7	3.5
3rd	7.1	4.4	9.0	5.6
4th	12.8	8.0	16.3	10.1
5th	27.2	16.9	34.5	21.4
Reverse	4.1	2.5	5.2	3.2

Torque convertor (Brockhouse)

Torque convertor stall ratio:	2.8:1
Gears:	Four

Maximum speed (standard tyres) with zero torque convertor slippage:

	Forward		Reverse	
	kph	mph	kph	mph
1st	4.5	2.8	4.4	2.7
2nd	7.1	4.4	7.1	4.4
3rd	12.8	8.0	12.7	7.9
4th	27.2	16.9	27.0	16.8

Powertrain (Build 5) variants

Manual synchromesh

Clutch:	Single dry plate type, 325mm (12.8in) diameter, 19,610mm (30in^2) friction area.
Gears:	Nine forward and three reverse speeds from three forward and one reverse speed full synchromesh gearbox, plus three-speed constant mesh range change, giving rough terrain (low), hard standing (medium) and roadway (high) ratios. Once the range has been selected, gears can easily be changed on the move.

Maximum speeds (standard tyres):

	Low		Medium		High	
	kph	mph	kph	mph	kph	mph
1st	3.1	2.0	7.2	4.5	17.1	10.6
2nd	4.0	2.5	9.1	5.7	21.7	13.5
3rd	5.4	3.4	12.3	7.7	29.4	18.2
4th	4.6	2.9	10.5	6.5	25.0	15.5

JCB Powertrain torque convertor

Componentised transmission, consisting of resiliently mounted engine, Borg Warner torque convertor, shuttle reverser and gearbox, connected via a propshaft to the JCB rear axle.

Torque convertor:	Stall ratio 2.2:1
Power shuttle reverser:	Hydraulically operated epicyclic reverser unit, with foot-operated off-load facility for gear changing and engine disconnect. Direction control by hand lever on the steering column.
Gearbox:	Four-speed full synchromesh unit for on-the-move gear change. Constant mesh gears on all speeds.
Axle:	JCB axle incorporating spur gear final reduction.

Maximum speed (standard tyres – ribbed earthmover):

	Forward		Reverse	
	kph	mph	kph	mph
1st	5.0	3.0	4.5	2.8
2nd	9.6	6.0	8.7	5.4
3rd	16.6	10.3	15.1	9.4
4th	27.2	16.9	24.7	15.4

Technical data

Electrical system	
Type	12V negative earth
Battery	12V at 128A/hr, at 20hr rating
Alternator	43A, rated output, sealed –37A
Up to engine 498NT 2724	Lucas 18ACR
From engine 4/98 2725	AC Delco DN460
Sealed version	CAV AC5HER
Starter motor	Lucas M50
Light bulbs	
Headlights	50/40W
Working lights	48W
Side lights	5W
Stop/tail lights	21/5W
Indicators	4V x 21W
Licence plate	2 x 5W
Instruments	18W
Interior	18W Festoon
Fuses	6 x 8A, 2 x 16A

Engine	
Make	BLMC four-cylinder
Type	4-98 NT direct injection
Swept volume	3.77 litres (230in^3)
Bore	98 to 98.2mm (3.8583 to 3.859in)
Compression ratio	16.8:1
Firing order	1, 3, 4, 2
Valve clearance (hot or cold)	0.33mm (0.013in)
Injection timing:	
Simms Minimec	16° BTDC
CAV DPA	14° BTDC
Engine speed:	
Idling	700–720rpm
Maximum revs	2,000rpm
Maximum no load revs	2,200–2,220rpm
Coolant temperature	70–95°C (158–203°F)
Gross power	52.2kW (70hp)
Net power	44.kW / 59.7hp
Peak torque @ 1,300rpm	248Nm (183lbf/ft)
Fuel consumption:	
at 50% maximum power, 1,300rpm	4.77 litres/hour (1.05gal/hr)
at 75% maximum power	7.15 litres/hour (1.5gal/hr)
Oil consumption:	
Skid variant (pre-Powertrain)	1% of fuel consumption
Powertrain	0.6% of fuel consumption

Hydraulics		
Pump flow rates at 2,000rpm and maximum pressure.		
Hamworthy gear – 126.1 litres per minute (33.3gal/min)		
Relief valve operating pressures:	bar	lbf/in^2
Main relief valve (MRV)	138	2,000
Auxiliary relief valve (ARV):		
Hamworthy (shovel and excavator)	186	2,700
Slew cylinders 127mm (5in) only	152	2,200
Cessna shovel dump	207	3,000
Cessna shovel clam	190	2,750
Slew torque	6,589Nm (4,860lbf/ft)	

Steering	
System type	Fully hydrostatic
Pump	Plessey gear type
Control unit	Danfoss Orbital
Maximum flow to control unit:	
Manual clutch machines	18 litres/min (4.8 US gal/min)
Build 3 to machine no 131319	36–38 litres/min (9.6–10.2 US gal/min)
Build 3 from machine no 131320	16.3–19.1 litres/min (4.3–5 US gal/min)
Relief valve operating pressure:	
To machine no 131319	121 bar (1,750 lbf/in^2)
From machine no 131320	117–124 bar (1,700–1,800lbf/in^2)
Camber Angle	5°
Castor Angle	2°
Toe-in	1.6mm (1/16in)

Tyres	Size (in)	bar	psi
Standard front:			
Vredestein	9.00 x 16, 10-ply	3.93	57
Other makes	9.00 x 16, 10-ply	3.44	50
Optional front	10.50 x 16, 12-ply	4.41	64
	10.50/80 x 18, 10-ply	4.62	67
	12.00 x 18, 10-ply	2.76	40
	12.00 x 18, 12-ply	3.93	57
	*13.0/65 x 18, 12-ply	3.59	52
Standard rear	16.9/14 x 28, 8-ply	1.65	24
Optional rear	16.9/14 x 28, 10-ply	2.07	30
	18.4/15 x 26, 8-ply	1.38	20
	18.4/15 x 26,-12-ply	2.14	31

* Specific tyres for machine options –
JCB Powerbreaker, extending dipper and Jaw Bucket:
   Front: As available for standard machine
   Rear: 18.4/15 x 26, 12-ply

Appendix F
Export standard builds specifications

Model	Power unit & transmission	Shovel	Tyres and wheels	Heater demister	Hydraclamp	TUV	Cab	Glass	Bostrom seat	Windscreen washer	Cab soundproof	Hazard warning
African	As standard	1¼yd³	9.00x16,10-ply ribbed 16.9/14x28, 8-ply traction	No	Yes	No	Option	Option	No	No No wipers	No	No
Australian	As standard	None	9.00x16,10-ply truck 16.9/14x28, 8-ply industrial	No	Yes	No	Std	No	No	No	No	No
Austrian	Austrian	1¼yd³ Tooth guard. Headlights inside bucket. Indicators on top.	9.00x16, 10-ply ribbed 18.4/15x26, 12-ply traction	Yes	Yes	Aus	Std plus Aust TUV	Std	No	Yes	No	No
French	Standard governed soundproof	1¼yd³	12.00x18,10-ply industrial 16.9/14x28, 8-ply traction	Yes	Yes	No	Std	Std	No	No	No	No
German	German	1¼yd³ Tooth guard. Indicators on top. Bracket to carry exc bucket.	9,00x16,10-ply ribbed 18.4/15x26,12-ply traction	Yes	Yes	Ger	Std plus TUV	Std	Yes	No	Yes	Yes
Gulf	As standard	1yd³ 6-in-1 clam shovel.	12.00/75x18, 10-ply industrial 18.4/15x26, 12-ply traction	No	Yes	No	Std	Std	No	No	No	No
Japanese	Japanese	None. Fit pins.	9.00x16, 10-ply ribbed 16.9/14x28, 8-ply traction	Yes	Yes	Jap	Shell only. Interior light	No	No	No	Roof lining only	No
Malaysian	As standard Dry battery	1½yd³	9.00x16, 10-ply ribbed Dunlop 16.9/14x28, 8-ply traction Dunlop	No	Yes	No	No. One wiper motor	No	No	No	No	No
Swiss	As standard	1½yd³ Tooth guard. Headlights & indicators on top of bucket.	12.00x18, 12-ply industrial 18.4/15x26, 12-ply traction	Yes	Yes	Swiss	Std plus Swiss TUV	Std	No	No	No	No
USA & Canada	As standard	11.4yd³	9.00x16x10 ply 16.9/14x28, 8-ply industrial Dunlop	Yes	Yes	No	Std safety Rops plate	Std	Yes	Yes	Yes	Yes

Useful contacts

Internet
J.C. Bamford Excavators
www.jcb.com

Web forums
Classic Machinery Network
www.classicmachinery.net

Facebook group
Vintage JCB machines
JCB models, merchandise and more.

YouTube
JCB 3C

Addresses

Abiljo Excavator Services Ltd
Fauld Industrial Estate
Fauld Nr Tutbury
Burton upon Trent
Staffordshire DE13 9HS
Tel 01283 815544
Website www.abiljo.com
Manufacturers of buckets and attachments.

Agriline Products Ltd
Unit 15, Harris Business Park
Hanbury Road
Stoke Prior
Bromsgrove
Worcestershire
B60 4DJ
Tel 01527 579111
Website www.agrilineproducts.com
Tractor spare parts.

Airfawn Consultants Ltd
Unit 4, Keates Way
Cheadle
Staffordshire ST10 1UR
Tel 01538 750788
Website www.airfawn.co.uk
Manufacturers of wiring harnesses.

Albert Jagger Ltd
Centaur Works
Green Lane
Walsall
West Midlands
WS2 8HG
Tel 01922 471000
Website www.albert-jagger.co.uk
Industrial hardware suppliers – cab, body, door and window fittings, fasteners, locks and latches.

Aldridge Trimming Ltd
Castle House
Drayton Street
Wolverhampton
West Midlands
WV2 4EF
Tel 01902 710805
Website www.aldridge.co.uk
Specialist vehicle trimming and upholstery.

Allmet Fabrications Ltd
Unit 3
Gaveston Grove
Oldfields Business Park
Fenton
Stoke-on-Trent
Staffordshire
ST4 3ES
Tel 01782 593222
Website www.allmetfabrications.co.uk
Metal folding and fabrications.

A&L Springs
Unit 2
29 Wigston Street
Countesthorpe
Leicester
LE8 5RP
Tel 0116 2788100
Website www.aandlsprings.co.uk
Custom spring manufacturers.

Border Plant Sales Ltd
Orchard Farm
Brookhouse Green
Smallwood Nr Sandbach
Cheshire CW11 2XE
Tel 01477 500305
Website www.borderplantsales.com
Tractor spare parts.

Brake Re-Line Services
Unit C2, West Point Industrial Estate
Penarth Road
Cardiff CF11 8JQ
Tel 029 20702900
Brake and clutch relining specialist.

Burton Rubber Co Ltd
Anglesey Road
Burton upon Trent
Staffordshire DE14 3NX
Tel 01283 567717
Website www.burtonrubberco.com
Manufacturers of mouldings, gaskets and extrusions.

Charles Leek & Sons Ltd
Springfields Works
Ashbourne Road, Leek
Staffordshire ST13 5AY
Tel 01538 382066
Website www.leakgears.co.uk
Precision gear cutting and machining specialist.

Diamond Metal Finishing Company Ltd
Unit 6, Newfields Industrial Estate
High Street
Stoke-on-Trent
Staffordshire ST6 5PD
Tel 01782 822442
Website
    www.diamondmetalfinishing.co.uk
Electroplating coating specialist.

Direct Hydraulics Services Ltd
Unit 2, Churnet Industrial Estate
Harrison Way
Cheddleton
Staffordshire
ST13 7EF
Tel 01538 361777
Website www.direct-group.co.uk
Hydraulic valve block and ram repair specialist.

Dynasurf (UK) Ltd
Millbuck Way
Springvale Industrial Estate
Sandbach
Cheshire
CW11 3GQ
Tel 01270 763032
Website www.dynasurf.co.uk
Hard chrome and precision cylindrical grinding (valves and spools).

Eric Warburton (Plant and Machinery) Ltd
Dairy House Farm
Ashley
Altrincham
Cheshire
WA15 0QG
Tel 0161 9281501
Website www.warburton-parts.com
Used JCB machines and parts.

Exilco Ltd
203 Foley Industrial Estate
Beauchamp Avenue
Kidderminster
Worcester
DY11 7DH
Tel 01562 823034
Website www.exilco.co.uk
Metal labels and nameplates.

Fairwayseals Ltd
Fernie Road
Market Harborough
Leicestershire
LE16 7FS
Tel 01858 410567
Website www.fairwayseals.co.uk
Hydraulic seal specialists.

First Foundations Ltd
Poplars Farm
Moor Lane
Ockbrook
Derbyshire
DE72 3SA
Tel 01332 670896
Website 1stfoundations.co.uk
JCB machines and parts.

Flexolite
Old Brickwood Farm
West Grimstead
Salisbury
Wiltshire
SP5 3RN
Tel 01722 712829
Website www.flexolite.co.uk
Custom-made fuel hoses and fittings for classic vehicles.

G. & S.J. Johnson Engineering Ltd
Parkhall Farm
Adderley Road
Cheadle
Staffordshire
ST10 2NJ
Tel 01538 756272
Website www.johnsonag.co.uk
Specialist engineering and agricultural engineers.

G.E. Middleton & Co Ltd
361/363 City Road
Manchester
ME15 4DE
Tel 0161 8720923
Website www.gemiddleton.com
Auto electrical component specialist.

GKN Driveline Services Ltd
Higher Woodcroft
Leek
Staffordshire
ST13 5QF
Tel 01538 384278
Website www.gknservice.com
Hydraulic ram manufacturers and repairs, propshaft manufacturers and refurbishment.

Hydrafit Ltd
Walsall Street
Wolverhampton
WV1 3LN
Tel 01902 451172
Website www.hydrafitlinwood.co.uk
Suppliers and manufacturers of hydraulic hoses, fittings and pumps.

Iracroft Ltd
Blandford Heights Industrial Estate
Shaftesbury Lane
Blandford Forum
Dorset
DT11 7TE
Tel 01258 486300
Website www.iracroft.co.uk
Hydraulic and coolant tube suppliers and specialist pipe benders.

J. Charnley & Sons
Marsh Lane
Brindle
Chorley
Lancashire
PR6 8NY
Tel 01254 854103
Website www.charnleys.com
Leyland parts and agricultural engineers.

JD Blasting and Painting Services Ltd
Adderley Mill
Brookhouse Road
Cheadle
Staffordshire
ST10 2NJ
Tel 07952 520850
Shot blasting and painting specialist.

JTP Ltd
Rope Street
Shelton New Road
Hartshill
Stoke-on-Trent
Staffordshire
ST4 6DJ
Tel 01782 711755
Website www.jtpnonferrous.co.uk
Non-ferrous metal stockholders.

Mayhill Tractors
Elmbridge Farm Buildings
Sterrys Lane
May Hill
Gloucestershire
GL17 0NF
Tel 01452 831820
Website www.mayhilltractors.com
Vintage tractor parts.

Punchards Haulage Ltd
Bramshall Industrial Estate
Bramshall
Uttoxeter
Staffordshire
ST14 8TD
Tel 01889 569493
Website
   www.punchardshaulage.co.uk
Specialist haulage in the UK and Europe, low-loaders and crane-equipped rigids up to 32 tonnes.

Potteries Diesel Ltd
Mason Street
Fenton
Stoke-on-Trent
Staffordshire
ST4 3NE
Tel 01782 744556
Diesel fuel injection and auto electrical specialists.

Pugh and Sanders Ltd
Cotton Brook Road
Derby
DE23 8YJ
Tel 01283 510824
Website www.pughandsanders.co.uk
Industrial fasteners, imperial nuts and bolts specialist.

Seals+Direct Ltd
Unit 6
Milton Business Centre
Wick Drive
New Milton
Hampshire
BH25 6RH
Tel 01425 617722
Website www.sealsdirect.co.uk
Manufacturers of rubber sections.

The British Rubber Company
The Old Station
Baildon
West Yorkshire
BD17 6SE
Tel 01274 585427
Website
   www.thebritishrubberco.co.uk
Tractor tyre specialist.

Truckfix Ltd
Highfield Works
Bradford Road
Cleckheaton
West Yorkshire
BD19 3TN
Tel 01274 878648
Website www.truckfixclutches.com
Specialist in clutch manufacture and restoration.

Vicary Plant Spares UK Ltd
The Old Station
Station Road
North Kilworth
Leicestershire
LE17 6HY
Tel 01858 880219
Website www.vicaryplant.co.uk
New and used JCB parts.

Vintage Brass Plates
Woodside
Seanor Lane
Lower Pilsley
Chesterfield
Derbyshire
S45 8DQ
Tel 07594 969146
Reproduction brass plates.

Woolies (I & C Woolstenholmes Ltd)
Whitley Way
Northfields Industrial Estate
Market Deeping
Peterborough
Cambridgeshire PE6 8AR
Tel 01778 347347
Website www.woolies-trim.co.uk
Classic vehicle trim and upholstery specialist.

Vehicle Wiring Products Ltd
9 Buxton Court
Manners Industrial Estate
Ilkeston
Derbyshire
DE7 8EF
Tel 0115 9305454
Website www.vehicleproducts.co.uk
Auto electrical components.

Vintage & Collectors Car Spares
Beltoft House
Belton Road
Beltoft
North Lincolnshire
DN9 1NE
Tel 01724 784230
Specialist supplier of suspension parts, track rod ends, engine components and head gaskets.

Index

Advertising/sales literature 16-17, 21, 25, 60, 116-120
Agricultural industry/farmers 6, 21, 28, 84
Attachments – see also Excavators and Loaders 137-145
   bucket types 139
Auctions 84-85
Axles 52-53, 85, 94, 99, 132, 135

Backhoe loaders 6-7, 30, 33-36, 46, 76
   Case 34, 36, 39, 76
   Ford 34-36, 117
   International 34
   John Deere 36
   Massey Ferguson 34, 36, 64, 117
   Steelfab 34, 36
   Whitlock 34, 36, 118
Bamford, Lord Anthony 12, 121
Bamford Cyril Joseph 10-11
Bamford, Henry 10
Bamford CBE, Joseph Cyril (Mr JCB) 10-11
Bamford, Samuel 10
Bamfords Ltd of Uttoxeter 10, 16, 116
   Leighton Iron Works 10
Bearings 107
Brakes 48-49, 83-84, 86, 132, 135
   adjustment 100
   footbrake 100
   parking 101, 105, 135
Brassington, Julia 10
Buckets – see Excavators
Buying checks 85-88

Cabs 22-23, 25-26, 29-39, 60-62, 90-91, 122, 133
   accessories 65
   access 61-62, 133
   air conditioning 69-70, 127
   Caribbean 61, 122
   carpet 127
   controls and levers 17, 22, 2, 29, 46, 51, 55, 57, 62-65, 136

   foot pedals 71
   throttle 133
   corrosion 84
   dashboard/console 60, 86, 91, 133, 136
   fibreglass 28
   glazing/windscreen 60-61, 84, 91, 122, 124, 127, 133-134
   hour clock 86
   instrumentation 65, 86, 133, 136
   open canopies 61
   Powertrain 60
   safety cab 61
   seats and driving position 24, 29-31, 63-64, 84, 136
      Bostrom 63
      suspension 63, 127
   soundproofing 65-66
   'Spaceview' 27
   steering wheels 60, 133
   tarpaulin cover 66-67
   toolbox 62
   warning lights 63
   White 60-61
Care and safety 92-93, 106
Charles, HRH Prince 124
Chaseside loading shovel 7
Chassis – see JCB Hydrachassis system
Colour schemes 20, 29, 41, 60, 63, 89, 91-92, 122, 126-127
Competitors 34-36, 117-118
   comparative testing 78
   seat designs 64
Contract Journal 26, 116
Construction Equipment Exhibition, Crystal Palace, 1961 25
Construction industry/builders/contractors 6, 20-21, 24. 26-27, 36, 79, 120
Construction News 117
Corrosion 84, 87, 90

Dancing Diggers troupe 9, 23
Dealer network 20, 36, 119

Earth moving companies 24-25, 120
Earth moving equipment 24
   Inchgreen 24
   International track loaders 26
   Massey Ferguson 24
   Steelfab 24
   Wain-Roy 24
   Whitlock 24
Electrical system 64-65, 91-92, 112
   alternator 64, 112
   batteries 29, 66
   charging circuit check 112
   components 92
Engine power units and skid units 41-45
   BLMC 44
   BMC/Leyland 26, 28, 30, 41
   BMC Nuffield 26, 28
   Fordson 23-26, 41
   JCB EcoMAX 127
   Leyland 4-98 skid unit 30, 39, 41, 44-45, 48
Engines 85, 131, 149
   air cleaner 45, 96-97, 131
   balancing unit 45
   bay 92
   bonnet 61, 129
   cold start 45, 69-70, 81
   cooling system 44-45, 86, 96
      anti-freeze 96
   early problems 41-42
   fan belt 96
   mounting 135-136
   protection/sump guard 77, 131
   serial numbers 146
   soundproofing 66
   speed adjustment 62
   wet cylinder liners 44
Excavator buckets (scoops) and mounting – see also Attachments and JCB backhoe 17, 41, 49-52, 62-63, 66, 75-77, 93, 130, 137-145
   backfill blade 142
   boom 49, 75-77, 111

clay spade 141
cutting angle and level
 indicators 145
design 76
dipper 49, 63, 75-77, 84, 111
 extending 70-71, 93, 130, 141, 148
ditch cleaning bucket 140
Jaw Bucket 63, 71, 84, 125, 130,
 138-140, 144
powerbreaker 142
performance 147
pipework 84
prices 128
rear frame 50
ripper tooth 142
rod end-damping 58-59
SAE bucket ratings 138
slew system 50-51, 80, 110, 129
square-hole bucket 140
tapered ditching bucket 141
tapered ejector bucket 141
tipping link 77
valve block 57-58, 87
Exhaust system 66, 131
Exports 33-36, 53, 119, 122, 150

Factory-fitted options 128
Fatigue life design 49, 76
Finney, Henry 24
Fuel system, fillers, filters and tanks 23,
 29, 40, 88, 94-95, 132
 bleeding 94-95
 cleaning 88
 injection pump 96
 lift pump 95
 sediment bowl 95

Hamworthy Hydraulics Ltd 22, 55
Harrison, Arthur 12, 14
Henley, Martyn 124
Hilary, Sir Edmund 16
Hirst, Bill 12, 14, 35
Holnes, Bert 12
Hooper, Noel 23
Hydraulic systems 6-7, 20-22, 24-25,
 50, 55-60, 67, 71, 79-80, 86, 102-
 103, 130-133, 149
 circuits 71, 121
 cylinders 58
 filtration 4, 59, 102-103, 130
 fluid 101-102, 130

hoses and pipework 52, 59, 67-68,
 80, 84, 87, 106
main relief valve (MRV) 113
main suction strainer 102-103
oil reservoirs and tanks 14, 22-23,
 29, 40, 59, 76
pump 56, 67, 87, 93, 130
rams 58, 71
tractor 28
J.C. Bamford Excavators Ltd
 (JCB) 15
 acquires Chaseside Engineering 7
 awards and bonuses given 20, 23,
  34-35
 Crakemarsh Hall factory, near
  Uttoxeter 12, 14
 demonstration ground,
  Ashbourne 88
 drawing offices 75, 81, 121
 efficient design 74, 78
 factories 126-127
 founding 11
 golf range 144
 HQ, Lakeside Works, Rocester 11,
  14-15, 123
 manufacturing efficiency 36, 58
 operator training school 13
 publicity department 116-117
 Queen's Award to Industry 35
 Wrexham transmission
  factory 134
 70th anniversary 126
JCB models and products
 backhoe loaders 33-35, 74, 92,
  126-127, 137
 grass-mowers 16, 116
 Half-Tracks 16
 High Speed Major Loader 14
 Hydrachassis system 21-22, 26,
  40-41, 75-76, 79, 88, 129
  structural failures 88
 Hydraclamp system 51, 63
 Hydra-Digga 21-23, 55, 92
 Hydraslide system 25-26, 50, 75
 hydraulic backhoe 16, 26, 29-31
 Hydraulic Excavator Trencher 16-17,
  119
 hydraulic tipping trailer 12-14
 JCB 1 'Grave Digger' 26-27, 30
 JCB 1D dumper 24. 26
 JCB 2 26-30

JCB 2B 28-29
JCB 2D 28; Mk II 28
JCB 3 24-27, 29-30, 35, 50, 54,
 117-118, 120-121, 124-125
JCB 3C 26, 28-31, 33, 46, 61, 64,
 89, 121, 129-139
 Mk II 19, 29-30, 36, 51, 53, 66-67,
  70-71, 76-77, 116, 122-123, 140,
  144, 146
 Mk III 4, 19, 30-31, 36, 39-71,
  83-103, 115-116, 119-127, 134,
  141, 146
JCB 3CX 31, 36, 89, 92, 115, 126-
 127
 Platinum limited edition
  126-127
JCB 3D 35, 124
JCB 3D 28, 30
JCB 4 22-24, 119
JCB 4C 24-25, 30, 118
JCB 4D 28-29
Jeeps refurbished 13
LiveLink telematics system 127
Loadall power shovel 14, 21, 23
Loadall 65 21-22
Loadall 75 21
Major Loader 13-15, 16, 22
 Mk I 20-21, 118
Master Loader 15, 21
MK I Excavator 20-21
Powertrain transmission models 60,
 62, 65, 84, 94, 97-99, 100-101,
 105, 125, 134-136
Project 200 31
Roadbreaker 67-71, 134, 140
 tools 67-69
road-sweeper collector 28
Si-draulic Loader 15-16
JCB Research 79-80
JCB Service's parts
 department 120

Kingposts and carriage 25-26, 28-29,
 50, 76, 111, 130
 clamping 130

Land Rovers 28
Lichfield Agricultural Works 10
Lights 64, 92
 headlamps 66, 103
 warning 63, 65

Loaders (shovels), arms and mounting 41, 51-52, 56, 62, 84-85, 87, 93, 129, 142-145
   bulldozer blade 145
   design 77-78
   fast dump 57
   'float' position 57
   fork lifts 129, 143
   jib crane 144
   performance 148
   pole-planting rig 28, 144
   self-levelling 57
   tooth guard 66
   valves 22, 56-57
   6-in-1 clam shovel 144-145
Logbooks 83, 85
Logos, badges and plates 15, 20, 115-116, 120, 127, 146
Lubricants 113

Machine numbering 146
Maintenance 93, 107
   body and framework 110
   cleaning components 106
   dismantling and assembly 106
      component removal 110-111
   greasing points 93-94
   torque settings 110-111
   troubleshooting 113
Merchandise 120
   clothing 120, 127
Mock-ups and models 79
Model years 31

Oil systems
   engine and gearbox 85, 94, 97-98
   rear axle 99
Optional and additional equipment 65-67

Paint 92
Parker Foundry Ltd, Derby 41
Pins and bushes 74-75, 87
Performance 49, 75-76, 145
   3C Mk III Excavator 147
   3C Mk II Loader 147
Power take-off 28
Plant hirers 6-7
Pre-delivery inspection (PDI) 81
Pricing 128

Production and sales figures 16, 22, 27, 29, 34-36, 126-127
   1,000,000 machines 126
Prototypes 79, 121
Public Works & Municipalities Exhibition, Olympia 1962 27

Restoration 84, 88-92, 123
   budget 89
   cleaning parts 92
   necessary skills 89, 91
   sheet metal work 90-91
   sourcing parts 89-90, 92
   storage site 89
   stripping down the machine 90
River Authority Open Day 1975 70
Road/rail machine 141
Road use 85
Rope shovels 6-7
Rotating flashing beacon 65
Royal Show 1953 17

Scale models and kits 120-125, 127
   digger in a bottle 123
   gold 122
Servicing 107
   serviceable parts 108
   service schedule 108-109
Shovels – see Loaders
Spare parts 27, 89-90
   part numbering 81, 119
Specifications 55, 119
   choosing 84
   export standard builds 150
   JCB 3C Mk III 147-149
   JCB 3C Mk III and 3CX comparison 127
   technical data 149
   weight 148
Spenborough Engineering Ltd 17
Stabilisers 50, 75, 141
   street pads 66
Steam power 7
Steering 52-55, 83, 86, 132, 149
   motor 55
   power-steering system 28-30, 54-55
   pump 55
   pump strainer 103
   ram 53-54
Story of JCB museum 73

S. Whiteley and Sons Ltd, Cleckheaton 61

Tachograph cable 131
Testing 76, 78-81
   axles 135
   electrics 112
   laboratory 78-81
   pressure 107
      main relief valve (MRV) 112
   transmission 135
Tickhill, Herbert 14
Torque and power 74
   definition of horsepower 74
Track loaders and machines 22
   County Crawlers 22
   International 22
   Track Marshall 22
Tractors 6, 15
   BMC Nuffield 22
   Ferguson 16-17, 21
   Fordson 20, 22
   W27N 15
   Major 13, 16-17, 21-22
   Morris Motors Nuffield Universal 13, 17
Transmission 31, 45-48, 86, 125, 131, 134-136, 146, 148
   automatic 127
   clutch 44-47, 99-100
   final drive 48
   gearboxes 31, 45-47, 97, 131
   how it works 47-48
   Powertrain 134-136, 148
   propshaft 94, 105
   protection 77
   serial numbers 146
   synchromesh gears 135
   torque convertors 39, 46-48, 84, 97-99, 132, 136, 148
      Brockhouse 46, 48, 84, 125, 148
      Lucas 46
Turner, Dolores 11
Tyres 27-28, 51, 53-54, 83, 124, 149
   types 53

Watt, James 74
Wheeldon, John 15, 35
Wheels 53-54
   front hubs 132